T0201820

Low-Rate Wireless Personal Area Networks

...Enabling Wireless Sensors with IEEE 802.15.4™

Third Edition

José A. Gutiérrez

Ludwig Winkel

Edgar H. Callaway, Jr.

Raymond L. Barrett, Jr.

Published by
Standards Information Network
IEEE Press

Trademarks and Disclaimers

IEEE believes the information in this publication is accurate as of its publication date; such information is subject to change without notice. IEEE is not responsible for any inadvertent errors.

Library of Congress Cataloging-in-Publication Data

Low-rate wireless personal area networks: enabling wireless sensors with IEEE 802.15.4 / José A. Gutiérrez ... [et al.]. -- 3rd ed.

p. cm. -- (IEEE standards wireless networks series)

Rev. ed. of: Low-rate wireless personal area networks / José A. Gutiérrez, Edgar H. Callaway, Jr., Raymond L. Barrett, Jr. 2007.

Includes bibliographical references and index.
ISBN 978-0-7381-6285-0 (alk. paper)
1. Wireless LANs. 2. Sensor networks. I. Gutiérrez, José A., 1967- Low-rate wireless personal area networks.

TK5105.78.G88 2010
004.6'8--dc22

2010037820

The Institute of Electrical and Electronics Engineers, Inc.
3 Park Avenue, New York, NY 10016-5997, USA

Samantha L. Bradley, Associate Product Manager
Marilyn G. Catis, Editor

Contact IEEE-SA

Review Policy

To order IEEE Press Publications, call 1-800-678-IEEE.

Print: ISBN 978-0-7381-6285-0   STDSP1180

See other IEEE standards and standards-related product listings at:
http://standards.ieee.org/

Trademarks

Bluetooth is a registered trademark of Bluetooth SIG, Inc. (www.bluetooth.org/).

DeviceNet is a trademark of ODVA (www.odva.org/).

Fieldbus is a registered trademark of the Fieldbus Foundation (www.fieldbus.org/).

HART and WirelessHART are registered trademarks of the HART Communication Foundation (http://www.hartcomm.org/).

IEEE and 802 are registered trademarks of the Institute of Electrical and Electronics Engineers, Incorporated (www.ieee.org/).

IEEE Standards designations are trademarks of the Institute of Electrical and Electronics Engineers, Incorporated (www.ieee.org/).

IrDA is a registered trademark of the Infrared Data Association (www.irda.org/).

PROFIBUS is a registered trademark of PROFIBUS Nutzerorganisation e.V (www.profibus.com/).

ZigBee Alliance is a trademark of the ZigBee Alliance (www.zigbee.org/).

All other trademarks belong to their respective owners.

Dedication

To my son Alan Alejandro, my newborn baby girl Emma Alessandra, and my wife Natacha, for their loving support and motivation to help me accomplish my dreams.

José A. Gutiérrez

To my wife Christel and my family (especially Anne, Dominik, Magnus, Bent, Katrin, and Roman) for their support and encouragement.

Ludwig Winkel

To my parents, Pat and Ed, Sr., for the example they set and the guidance they gave. And to my wife, Jan; without her support and understanding, this book would not have been possible.

Ed Callaway

To my wife, Julie, and our two wonderful children, Gloria and Raymond, who bring meaning to my life and work.

Ray Barrett

Acknowledgment

Behind this book is the work of the larger IEEE 802® community, in particular, the outstanding work performed by the technical editorial team of the IEEE 802.15.4™ Task Group, led by the indomitable and indefatigable Dr. Bob Heile.

We would like to express our sincere thanks to a diverse group of persons who provided their support during the standardization process, and with whom we have had many enriching technical discussions: Anthony Allen, Venkat Bahl, Daniel Bailey, Monique Bourgeois (IEEE Std 802.15.4 MAC Technical Editor), Priscilla Chen, Dr. Neiyer Correal, David Cypher, Ralph D'Souza, Dr. Larry Dworsky, Paul Gorday, David Hayes, Dr. Lance Hester, Dr. Ivan Howitt, Dr. Jian Huang, Dr. Yan Huang, Phil Jamieson (IEEE Std 802.15.4 Vice Chair), Pat Kinney (IEEE Std 802.15.4 Chair), Masahiro Maeda, Dr. Fred Martin, Dave McClanahan, Vinay Mitter, Said Moridi (IEEE Std 802.15.4 PHY Technical Editor), Dr. Bob O'Dea, Luis Pereira, Dr. Robert Poor, Larry Puhl, Phil Rudland, Dr. Chip Shanley, Qicai Shi, Bob Stengel, Rene Struik, David Taubenheim, and Hans van Leeuwen. In addition to many of the above, Farron Dacus, Barbara Doutre, Latonia Gordon, Charles Luebke, and Marco Naeve (IEEE Std 802.15.4 Secretary) commented on the many early drafts of this book.

We also would like to thank the anonymous peer reviewers of this book for their evaluation and guidance in helping to enrich the material presented here.

Finally, we would like to thank the IEEE editorial team for their diligence and great work.

Third Edition Acknowledgments

The creation of the third edition of this book is tightly linked to the creation of the WirelessHART standard and the team involved in making it happens. We wish to thank Dr. Kris Pister, the pioneer behind the core design of "Smart Dust," from which formed the technical foundation for WirelessHART. We also wish to thank Ron Helson, Bob Karschnia, Terry Krouth, Hans-Georg Kumpfmüller for creating and leading the vision for industrial wireless sensor and actuation networks for the process industry.

Many thanks to the expert leadership and guidance of the HCF Chief Engineer Wally Pratt, who served as the primary architect and lead designer of the WirelessHART technology and the team of technical experts whose great contribution and dedication was instrumental to the successful development and further deployment of the WirelessHART in the marketplace: Deji Chen, Rick Enns, Stefan Gross, Mark Nixon, Robin Pramanik, Eric Rotvold, and Gareth Johnson.

Furthermore, this book was possible only with the support of a large community of friends and technology enthusiasts: Venkat Bahl, John Berra, Leonel Bertuso, Walt Boyes, Steve Burke, Jim Cobb, Sergio Fernandes, Ashok Gupta, Chris Herzog, Sean Keeping, Mark Kennedy, Jane Lansing, Randall Ledford, Klaus-Peter Lindner, Mike Medley, Kelly Orth, Jeff Potter, Mark Schumacher, Paul Sereiko, Guido Stephan, Werner Thoren, Marty Zielinsky, Peter Zornio.

Finally, we would like to thank Samantha Bradley, our anonymous peer reviewers, and the IEEE SIN team for their outstanding work in making this book happen.

Authors

José A. Gutiérrez is Corporate Director of Technology at Emerson Electric, leading a wide range of technology activities that include the use of wireless sensor networks in the industrial environment. He received the B.S. degree in electronic engineering from Universidad Simón Bolivar in Caracas, Venezuela, in 1991, M.S. in electrical engineering from the University of Milwaukee - Wisconsin in 2001, and the Ph.D. degree from the same university in 2005. Through his active membership at the IEEE LAN/MAN Standards committee, in
2000 Dr. Gutiérrez became the Editor in-Chief of the IEEE 802.15™ Working Group, Task Group 4, focused in the development of low-rate wireless personal area networks. After the success of IEEE Std 802.15.4, he supported the creation of the ZigBee Alliance, where from 2003 to early 2005 he worked as program manager and from 2009, a member of its board of directors. Currently, Dr. Gutiérrez is the WirelessHART liaison between the HART Communications Foundation and IEC, IEC U.S. National Expert for IEC SC65C WG-17 focused on Wireless Coexistence, and technology advisor for these organizations. Dr. Gutiérrez has multiple patents in the area of wireless personal and local area networks and author of several of papers in the areas of automatic control, artificial intelligence, and wireless communications. Dr. Gutiérrez is also the author of a book on robotic vision, *Precision Landmark Location for Machine Vision and Photogrammetry: Finding and Achieving the Maximum Possible Accuracy* (Springer 2007). Dr. Gutiérrez is a Senior Member of the IEEE and active member of the IEEE Standards Association.

Ludwig Winkel received his degree "Diplom Ingenieur" (Dipl. Ing.), in electrotechnical, electronic and control engineering from the University of Saarbruecken, Germany, in 1977. He joint Siemens AG in Karlsruhe, Germany, Nov. 1977 in a R&D group for electrical measuring devices. Ludwig introduced micro controller technology and large scale integrated chips in the measuring devices that were revolutionary in 1977. This work lead to other R&D positions, e.g., that he wrote a operating systems as a basis to implement test

and measurement equipment. He became the position of an R&D group leader with the projects to invent closed loop control hardware and software for programmable logic controllers (SIMATIC® PLC), and led later on the maintenance group for the DCS system TELEPERM®. Ludwig contributed to the development of a new DCS system SIMATIC PLC7 and has written two books on the system design, *SIMATIC Software – Process control System PCS 7 – Validation support manual I* and *SIMATIC Software – Process control System PCS 7– Validation support manual II*.

Since 1992 Ludwig has been actively involved in development and promotion of standards for instrumentation and control systems. He became a member of the Interoperable systems project (ISP), and the IEC Subcommittee 65C for Function Blocks for process control and the fieldbus committee IEC SC65C WG6. In 1999 he joined the strategic group in Siemens Automation division for fieldbus communications and in this position has actively participated in standards development as convenor of the wireless coexistence working group in IEC SC65C WG17, the convenor of the fieldbus Maintenance team IEC SC65C MT9, the convenor of the Real-time Ethernet Working Group in IEC SC 65C WG11, the editor of the WirelessHART specification in IEC SC65C W16 (project IEC 62591), the editor of Electronic Device Description Language (EDDL) Working Group in IEC SC65E WG7, the editor of the "Industrial automation systems and integration — Open systems application integration framework — Part 4: Reference description for Ethernet-based control systems Amendment 1: PROFINET profiles" in ISO TC184 SC5 WG5, Co-chair of ISA 104 dealing with EDDL and the Liaison officer to IEC SC65E WG7, one of the editors of IEEE 802.15.4e, the rapporteur of ETSI ERM TG28 project TR102889-2, the vice chair of the German national committee for fieldbusses DKE K956, and the expert in various national and international committees.

Edgar H. Callaway, Jr. received the B.S. degree in mathematics and the M.S.E.E. degree from the University of Florida in 1979 and 1983, respectively, the M.B.A. degree from Nova (now Nova-Southeastern) University in 1987, and the Ph.D. degree in computer engineering from Florida Atlantic University in 2002. He joined Motorola in 1984 as an RF engineer working on trunked radio products, and in 1990 transferred to Motorola's Paging Products Group, where he designed

paging receiver and transceiver systems and was the lead receiver designer of Motorola's paging platform. In 2000 Dr. Callaway joined Motorola Labs, where he investigated low-power wireless networks and played a major role in the development of the IEEE 802.15.4 Low-Rate Wireless Personal Area Network standard. He was also Motorola's representative on the ZigBee Alliance Board of Directors, and Chair of the ZigBee Security Working Group. In 2009 Dr. Callaway co-founded Sunrise Micro Devices, specializing in low-voltage, low-power wireless systems. He is a Registered Professional Engineer (Florida), is the author of several books, book chapters, and papers, and has more than 40 issued U.S. patents. Dr. Callaway was recently named an IEEE Fellow, "for contributions to wireless sensor networks and low power design techniques for communications devices and systems."

Raymond Louis Barrett, Jr. received the B.S.E.E. degree from Case Institute of Technology (now Case Western Reserve University) in 1966, the M.S. in computer science and M.B.A. degrees from Nova (now Nova Southeastern) University in 1983 and 1982, respectively, and the Ph.D. in electrical engineering from Florida Atlantic University (FAU) in 1990. He taught as an Assistant Professor for Nova University and FAU, and as an Associate Professor at the University of North Florida from 1982 through 2001.
Concurrently with teaching at FAU, he joined Motorola's Paging Products Group in 1991, whereupon he initially designed various low-power paging receiver circuits. In 1991 he joined Motorola Labs as a Principal Member of the Technical Staff, where his interests included the development of low-power wireless components. He is presently the CEO of American Research & Development, LLC, offering Professional Engineering (PE) services. He is the author of eleven Continuing Education Unit (CEU) courses offered by SunCam to meet the requirements for PE license renewal in every U.S state. He is an Affiliate Professor at the University of New Hampshire (UNH) following his interests in VLSI design. He is a registered Professional Engineer (Florida) and has thirty-five issued U.S. patents and several published papers.

Foreword

There is a program offered by the IEEE Standards Association called IEEE Get 802®. Through this program, IEEE 802® standards are available to the public at no cost six months after publication. Within two years of its initial publication in 2003, IEEE Standard 802.15.4™ became the third most downloaded 802 standard behind IEEE Std 802.11™ and IEEE Std 802.3™. On the ZigBee™ Alliance web site you can download the ZigBee specification, which is built on 802.15.4. Downloads have been averaging a steady 500 per week. Clearly interest in 802.15.4 is intense. I was delighted when the first addition of Low-Rate Wireless Personal Area Networks hit the market. Having a concise, clearly written guide is invaluable to helping developers understand a lengthy standards document. Just as important is helping the market understand the application space and practical issues surrounding real world deployment. This book does it all, and the second edition brings us up to date on a rapidly evolving and growing market.

The authors of this book are uniquely qualified for the task because not only are they intelligent, witty, and articulate, they are inextricably bound to the creation of the standard in the first place. IEEE 802.15 as a working group was started with the initial thought of developing a standard for a low data rate class of radio. Around the same time, the Bluetooth SIG was announced and for a while the marketplace was confused as to whether Bluetooth would address the needs of sensor and control networks. By late 2000, it was clear that while Bluetooth was very effective for wireless headsets, it was not the solution for sensor networks which needed much lower cost, very long battery life, and support for mesh networking. The continuing need for a standard in this space led to the formation of a Task Group in 802.15 to draft what became IEEE Std 802.15.4.

I had the honor of chairing the initial activity, where José Gutiérrez learned a valuable lesson on how one gets volunteered in spite of oneself, and he became the group's technical editor. This gives him a unique perspective on every aspect of the standard. Additionally, Ed Callaway, Ray Barrett, Venkat Bahl, and Kursat Kimyacioglu all had significant background in low data rate radios and mesh networking. While there were many contributors, these gentlemen in particular worked tirelessly to merge this understanding into what became the technology baseline for 802.15.4. I can not imagine a better threesome than José, Ed, and Ray to create a useful guide and keep it up to date. In addition, Ed and José both contributed to the formation of the ZigBee Alliance. José served as the head of the Alliance Program Management Office, and Ed was on the Alliance Board of Directors. All of them remain active to this day in various aspects related to the deployment and enhancement of the technology.

Since the first addition of Low-Rate Wireless Personal Area Networks, much has happened. We launched a Physical Layer amendment called 802.15.4a™, which is in the final stages of balloting in the IEEE Standards process. We also completed a revision to the standard, which was published this year as IEEE Std 802.15.4™-2006. This revision cleaned up some issues in the MAC, substantially improved the security suite, and added higher data rate modulation modes to the sub-1 GHz bands. We are also about to start two new amendments, 802.15.4c™ and 802.15.4d™, to add sub-1 GHz band coverage in China and Japan, in addition to those currently supported in the USA and Europe. Also since the first addition, the ZigBee Alliance has published its specification, which adds the networking stack and application layer to 802.15.4. The ZigBee spec had been downloaded over 35,000 times as of October 2006. The bottom line of all this activity is that 802.15.4 is likely to become the most widely deployed radio on the planet over the next decade, because it will wind up in so many things.

Whether you are trying to get a first pass feel of what the market is all about or whether you are trying to get a better intuitive grasp of the standard and how it works, this is the place to come. Thanks guys, and if I do my job right, you better start making plans for the third edition.

Bob Heile
Chair, IEEE 802.15
Chairman, ZigBee Alliance

Foreword to the Third Edition

Since the release of the initial IEEE 802.15.4 standard in 2003, it has catalyzed a fire storm of activity, research, and interest in wireless sensor networks and prospective applications. The low cost, low power, commercial off-the-shelf radio chip sets and components enabled by IEEE Std 802.15.4 have been a game changer and clearly establish IEEE Std 802.15.4 as the core technology for wireless sensor networks.

I am excited about this book for several reasons. I have known the authors for many years and both are very knowledgeable experts in the field. I am particularly pleased that they chose WirelessHART® as the key feature of the third edition of this book. The authors very articulately show the progression of IEEE Std 802.15.4, presenting WirelessHART as a successful work product example based on this technology. This book fills a gap in the available literature, will help educate the community about these emerging standards, and will help grow the ecosystem for wireless sensor networks.

The advancements in wireless sensor networks enabled by IEEE Std 802.15.4 were instrumental in the HART® Communication Foundation decision to consider extending our industry standard communication technology to include wireless capabilities in late 2004. By early 2005, our wireless working group was fully engaged with the goal to develop a wireless communication standard for industrial process automation. In September 2007, we were delighted to achieve the goal with release of the WirelessHART standard specifically designed to meet the needs of industry for simple, reliable, and secure wireless communication in process measurement and control applications. In March 2010, WirelessHART became a full international standard approved by the International Electrotechnical Commission as IEC 62591.

WirelessHART (IEC 62591) is the only international standard for industrial wireless sensor networks based on IEEE Std 802.15.4 technology. The requirements for communication standards and systems in industrial process automation applications demand extremely high levels of reliability, security, and robustness. WirelessHART fully leverages IEEE Std 802.15.4 to meet these process industry needs and is changing the game for industrial wireless communication systems with unparalleled levels of reliability and security.

Needless to say, we are excited about the WirelessHART standard and the opportunities, possibilities and benefits this new technology brings to our industry. The development of this international industrial wireless communication standard would not have been possible without the hard work, tenacity and expertise of the many people and companies that contributed to its development and progress.

Before closing, I believe it important that to say a couple of things: WirelessHART is not for everyone. The focus of WirelessHART is industrial process measurement and control. We did not design it as a general purpose wireless sensor network for all applications. Also, WirelessHART does not compete with ZigBee. These wireless communication standards and technologies are both very important to the industry, but their market focus and performance metrics are very different.

Regardless of whether you are new to the field of wireless sensor networks or just looking to better understand the progression of IEEE Std 802.15.4 and what may be possible in your applications, this book is for you. The authors have done an excellent job, and my thanks go to them for allowing me to make a small contribution.

Ron Helson
Executive Director,
HART Communication Foundation

Table of Contents

Introduction and Update for the Third Edition

IEEE Std 802.15.4™ is the first and only wireless radio technology designed uniquely to enable communications between machines or machines and humans, covering residential, commercial, and industrial applications. IEEE Std 802.15.4 allows the creation of wireless sensing and actuation systems to enable machine-to-machine (M2M) communication systems, also called simply *wireless sensor networks* (WSNs).

Wireless sensor networks have several unique requirements that set them apart from conventional wireless communication technologies. Most wireless sensor applications are particularly constrained by the power consumption; technical performance requirements such as throughput, latency, reliability, and security; and other challenges such as cost of ownership and ease of use. IEEE 802.15.4 technology addresses each of these issues by defining a simple yet robust radio technology that can be configured via upper layer communication protocols, to meet the requirements of specific application spaces. This is the case, for example, of the ZigBee™ technology for residential and commercial applications and WirelessHART™ for industrial applications.

Since the publication of the original IEEE Std 802.15.4 in 2003, the interest in low-rate wireless personal area networks (LR-WPAN) and wireless sensor networks has increased exponentially. The wireless sensing market has generated an ecosystem of talented people, companies, systems, and tools to support applications focused on solving specific problems. These applications cover Smart Grid's home area networks, industrial field instrumentation connectivity, communication on consumer electronic devices, healthcare monitoring, condition-based monitoring in commercial environments, and several others.

As originally designed, IEEE Std 802.15.4 has become the key enabling technology for machine-to-machine communication applications and the benchmark of other technologies that want to enter this application space. Some examples of this dynamic are the new low-power Bluetooth® efforts (formerly called WiBree) currently in development and the newly introduced low-power 802.11™ systems that enable the use of wireless monitoring systems using the existing WiFi™ infrastructure. These technologies complement the wireless sensing landscape for specific application niches. However, they cannot compete under the same conditions and assumptions with the low-power consumption provided by IEEE 802.15.4 systems. The reason is the simplicity of IEEE 802.15.4 design. Simplicity allows for a simple and streamlined implementation. Simplicity in the semiconductor world also implies fewer gates, less code, and less processing power, all together supporting a much lower power consumption providing one or two orders of magnitude lower than other existing technologies. Of course, this low-power consumption has a trade-off in the form of lower throughput (low-rate), shorter propagation distance, and reduced features, as will be explained later in the book.

Another relevant dynamic comes from the shifting of proprietary wireless communication systems toward the use of the IEEE 802.15.4 radio. These systems can use IEEE 802.15.4 off-the-shelf radios, sourced from more than a dozen suppliers with a rich feature set while using the knowledge base of the IEEE Std 802.15.4 community.

IEEE STD 802.15.4 EVOLUTION

Following the initial release in 2003, IEEE Std 802.15.4 has continued evolving to include technical features that enable its use in new specialized applications such as location awareness, improve and enhance its use in new markets with the added support for the newly introduced wireless sensor network bands in China and Japan, and allows easier implementation such as the incorporation of a flexible time division multiplexing (TDMA), and time-slotted channel hopping (TSCH) for industrial applications. The following figure shows a pictorial of the evolution of IEEE Std 802.15.4.

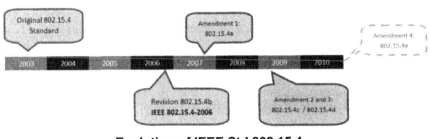

Evolution of IEEE Std 802.15.4

Revision IEEE Std 802.15.4 - 2006 (IEEE Std 802.15.4b)

In June 2006, the IEEE Standards Association approved a revision of IEEE Std 802.15.4 originally approved in 2003. The IEEE 802.15.4 Task Group 4b created this revision with the goal of removing ambiguities in the standard, while making technical improvements to the physical layer and medium access control layers. The PHY layer definition in this revision was expanded to include new optional modulation schemes:

- An 868/915 MHz DSSS PHY employing offset quadrature phase shift-keying (O-QPSK) Modulation.

- An 868/915 MHz parallel sequence spread spectrum (PSSS) PHY employing binary phase shift-keying (BPSK) and amplitude shift-keying (ASK) modulation

These new physical layers offer higher data rates in the 868/915 MHz bands, while adding more channels.

Amendment 1 – IEEE Std 802.15.4a

IEEE Std 802.15.4a, approved in August 2007, augments the IEEE Std 802.15.4 family by adding optional capability supporting precision ranging and location, as well as a higher aggregate throughput. This improvement was accomplished by adding two new optional physical layers:

- Ultra-wide band (UWB) PHY at frequencies of 3 GHz to 5 GHz, 6 GHz to 10 GHz, and less than 1 GHz. This physical layer supports an over-the-air mandatory data rate of 851 kb/s with optional data rates of 110 kb/s, 6.81 Mb/s, and 27.24 Mb/s.

- Chirp spread spectrum (CSS) PHY at 2450 MHz. The CSS PHY supports an over-the-air data rates of 250kb/s and 1000 kb/s.

The precision ranging capability was designed to be accurate down to one meter or better.

Amendment 2 – IEEE Std 802.15.4c

This amendment, approved in March 2009, adds two optional physical layers designed specifically to address the newly open 779–787 MHz Wireless Sensor Network band in China (also known as the 780 MHz Chinese band):

- MPSK PHY defined for the Chinese standard for the Wireless Medium Access Control (MAC)

- O-QPSK PHY specifications for Low-Rate Wireless Personal Area Networks

Amendment 3 – IEEE Std 802.15.4d

This amendment, also approved 19 March 2009, specifies two alternate physical layers designed to operate in the 950 MHz Japanese band:

- A direct sequence spread spectrum (DSSS) PHY employing BPSK modulation
- A PHY using Gaussian frequency-shift keying (GFSK) modulation

Amendment 4 – IEEE Std 802.15.4e

The IEEE 802.15 Task Group 4e is chartered with the assignment to define a MAC amendment to the existing standard 802.15.4-2006. The goal of this amendment is to enhance and add functionality to address the requirements of the industrial market. In particular it defines MAC level support for Time Division Multiplexing (TDMA) and channel hopping, which are key parameters to augment the wireless reliability in industrial facilities. A detailed technical explanation of the IEEE Std 802.15.4e functionality is presented in Part IV.

As of the first quarter of 2010, the IEEE 802.15.4e standardization effort is still a work in progress.

Amendment 5 – IEEE Std 802.15.4f

The IEEE 802.15 Task Group 4f is chartered to define new wireless Physical (PHY) layer and corresponding updates to the IEEE Std 802.15.4-2006 MAC layer required to support Active RFID System bidirectional and location determination applications. As of the third quarter of 2010, this amendment is work in progress.

Amendment 6 – IEEE Std 802.15.4g

IEEE Std 802.15.4g intends to define enhancements to create large and scalable outdoors networks for Smart Grid applications. As of the third quarter of 2010, this amendment is in a very early stage of development.

ZIGBEE AND MORE

The ZigBee Alliance was in most part created by the same people who authored the original IEEE Std 802.15.4. The goal of this group was to augment IEEE Std 802.15.4 with the upper communication layers needed to define a full protocol (following the 7-Layer ISO/OSI communications layering model) and provide marketing support for its success in the commercial and residential market place. However, today many other communication-based associations use IEEE Std 802.15.4 for a variety of application spaces. Examples are the Internet Protocol for Smart Objects (IPSO) Alliance, promoting the use of the IETF 6LoWPAN protocol, and the HART Communications Foundation, promoting the use an industrial-grade wireless communications protocol named WirelessHART; Part IV of this book covers WirelessHART technology in detail.

The third edition of this book extends the previous editions by adding a new section centered on providing a complete presentation of the WirelessHART protocol. This is the only existing international standard designed to provide very high reliability and performance for industrial sensing and actuation applications. This type of application represents the highest level of performance required for IEEE Std 802.15.4 because it involves extremely low power consumption, guaranteed throughput and latency, and highest level of security. As such, this new section provides more insights on how IEEE Std 802.15.4 can perform for this type of critical performance use case.

BOOK ORGANIZATION

This book is designed to complement IEEE Std 802.15.4 by presenting an overview of the features that characterize it, the applications that motivate it, and the rationale that dictated several of the design options chosen during its creation. The standard is written and structured in a way that makes it very easy to understand and (the authors believe) easy to implement. The IEEE Std 802.15.4 technical editing team concentrated especially on creating informative clauses to improve the communication of the more complex concepts found in the standard. This book does not pretend to extend that effort except in discussion of applications, but rather intends to be a companion work. It is directed to people interested in the field of "simple" wireless connectivity. It focuses on wireless sensors and actuators for the industry in general.

The material in this book is divided into four parts. Part I presents an overview of the low-rate wireless personal area technology and IEEE Std 802.15.4. The material focuses on the motivation for the creation of this standard, including the application scenarios that drove the making of it. Part I is designed not only as a technical introduction to the standard, but also to offer marketing and business professionals enough background information on the technology and the vision behind its conception to help plan marketing and business strategies.

Part II concentrates on the technical features and components of IEEE Std 802.15.4. It adds material on network layer functionality that is not covered by the standard. The network layer information applies to the rationale behind the design of several of the components of the standard for the envisioned applications. Part III focuses on implementation and system design considerations. It includes an analysis of system-level real-world issues that will be important for prospective implementers. Finally, Part IV introduces the WirelessHART communications standard, providing details showing how to build a high-performance and reliable industrial standard on the IEEE Std 802.15.4 technology.

The information in IEEE Std 802.15.4 is compact, and simplicity is reflected in the text. The standard is organized into seven clauses and seven annexes. Clauses 1 though 5 contain introductory material that maps with Part I of this book. Clauses 6 and 7 in the standard define the physical layer and medium access control (MAC) sublayer, respectively; these clauses map with Part II. Finally, as explained previously, Parts III and IV do not map to the standard, because it deals with implementation, usage, and upper layer protocols that the standard does not address.

Acronyms and Abbreviations

ACK   acknowledgment

ACL   access control list

AES   advanced encryption standard

ASK   amplitude shift-keying

BPSK   binary phase shift-keying

BSP   beacon synchronization parameter

CAP   contention access period

CBC-MAC   cipher block chaining message authentication code

CCA   clear channel assessment

CCM   counter mode + CBC-MAC

CFP   contention-free period

CRC   cyclic redundancy check

CSMA-CA   carrier sense multiple access with collision avoidance

CSS   chirp spread spectrum

CTR   counter mode

DLL   data link layer

DSSS   direct sequence spread spectrum

CWPAN   Chinese Wireless Personal Area Network

DPDU   Data link protocol data unit

ED	energy detection
ETSI	European Telecommunications Standards Institute
EVM	error-vector magnitude
FCC	Federal Communications Commission
FCS	frame check sequence
FFD	full function device
FHSS	frequency hopping spread spectrum
FM	frequency modulation
GFSK	Gaussian frequency-shift keying
GSM	global system for mobile communication
GTS	guaranteed time slot
HART	highway addressable remote transducer
HCF	HART Communications Foundation
HVAC	heating, ventilation, and air conditioning
IC	integrated circuit
IrDA	infrared data association
ISM	industrial, scientific, and medical
ITU-T	International Telecommunication Union - Telecommunication Services
LAN	local area network
LBT	listen before talk
LC	inductor-capacitor
LLC	logical link control

LQI	link quality indication
LR-WPAN	low-rate wireless personal area network
LSB	least significant bit
M2M	machine to machine
MAC	medium access control
MCPS-SAP	MAC common part sublayer service access point
MFR	MAC footer
MHR	MAC header
MIC	message integrity code
MIPS	million instructions per second
MLME	MAC sublayer management entity
MLME-SAP	MAC sublayer management entity service access point
MSB	most significant bit
MPDU	MAC protocol data unit
MPSK	M-ary phase shift-keying
MSDU	MAC service data unit
MSK	minimum shift-keying
Network Graph	abstract representation of network addresses in a mesh route
OEM	original equipment manufacturer
O-QPSK	offset quadrature phase shift-keying
OSI	open systems interconnection
PAN	personal area network

PD-SAP   PHY layer data service access point

PHY   physical

PHR   PHY header

PIB   PAN information base

PLME   PHY layer management entity

PLME-SAP   PHY layer management entity service access point

PN   pseudo-noise

POS   personal operating space

PPDU   PHY protocol data unit

ppm   parts per million

PSDU   PHY service data unit

PSSS   parallel sequence spread spectrum

QoS   quality of service

RAM   random access memory

RC   resistor-capacitor

RF   radio frequency

RFD   reduced function device

RFID   radio frequency identification

RSSI   received signal strength indication

RX   receive or receiver

SAP   service access point

SAW   surface acoustic wave

SHR   synchronization header

SNR   signal-to-noise ratio

SoC   system-on-a-chip

SSCS   service specific convergence sublayer

TDMA   time division multiple access

TSCH   time slotted channel hopping

TX   transmit or transmitter

VCO   voltage controlled oscillator

UWB   ultra-wide band

WiFi   wireless fidelity

WLAN   wireless local area network

WPAN   wireless personal area network

WSN   wireless sensor network

Part I

Chapter 1   IEEE Standard 802.15.4

...enabling wireless sensor networks

The explosive growth of embedded control and monitoring in almost any electronic device and the need for connectivity of these applications is causing an integration bottleneck. Manufacturers use different communication interfaces— standard or proprietary—creating the need for yet another controller task to make the arbitration. Conventionally, these communication links are wired. Wires allow power and the reliable transmission of signals from a controller to its peripherals. When the peripherals are not physically contained in the controller, the required wiring brings issues such as cost of installation, safety, and operation convenience to the surface. Wireless technology is the obvious solution to overcome these obstacles, although it comes with its own set of challenges—propagation, interference, security, regulations, and others. The technology to overcome these issues exists, but normally with added complexity, causing an increase in the cost of the system.

Certainly some applications can afford the cost of adding a high-end wireless communication system such as broadband data services from mobile phones, IEEE Std 802.11™ Wireless Local Area Networks, IEEE Std 802.16™ Wireless Metropolitan Area Networks, and so on. Conversely, many other applications can be enhanced or enabled with the use of standards based low-cost wireless communications solution made available by IEEE 802.15.4 technology. Standards drive interoperability among different manufacturers, enhance economy of scale, accelerate maturity, and minimize risk to the product developer with direct benefit to the end consumer.

An IEEE 802.15.4 low-rate wireless personal area network (LR-WPAN) is a network designed for low-cost and very low-power short-range wireless communications among machines or between machines. This definition is at odds with the current trend in wireless technologies whose focus has been on communications with high data throughput and enhanced quality of service (QoS) to support multimedia applications.

The current trend in wireless technologies is to produce protocols with higher data throughput and QoS as their primary performance metrics, with cost and power consumption secondary. The primary performance metrics of an LR-WPAN are cost and power consumption, with relaxed data throughput.

WLANS, WPANS, AND LR-WPANS

Wireless Local Area Networks (WLANs) were created as the wireless extension of the IEEE 802® wired local area network (LAN), which was designed for high-end data networking. Among the system requirements of a WLAN are seamless roaming, message forwarding, longest possible range and capacity for a large population of devices. In contrast, WPANs are designed to function in the personal operating space (POS), covering the area around a person, whether stationary or in motion.

WPANs are used to convey information over relatively short distances among the participant transceivers. Unlike WLANs, connections affected by WPANs involve little or no infrastructure. This allows small, power efficient, inexpensive solutions to be implemented for a wide range of devices.

The IEEE 802.15 Working Group has defined three classes of WPANs that are differentiated by data rate, battery drain, and QoS. The high-data rate WPAN (IEEE Std 802.15.3™) is suitable for multimedia applications that require very high QoS. Medium-rate WPANs (IEEE Std 802.15.1™/Bluetooth) are designed as cable replacements for consumer electronic devices centered on mobile phones and PDAs with a QoS suitable for voice applications. The last class of WPAN, LR-WPAN (IEEE Std 802.15.4), is intended to serve applications enabled only by the low power and cost requirements not targeted in the previous WPANs. LR-WPAN applications have a relaxed need for data rate and QoS. Figure 1–1 illustrates the operating space of the IEEE 802 WMAN, WLAN, and WPAN standards. Notice that IEEE Std 802.15.4 is not designed to overlap with higher end wireless networking standards.

Table 1–1 presents a summary of the principal characteristics of an LR-WPAN using IEEE Std 802.15.4, compared with IEEE Std 802.11b/g and a standard WPAN such as IEEE Std 802.15.1™.

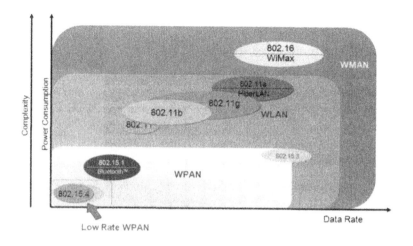

Low Rate WPAN

Figure 1–1: Operating space of WMAN, WLAN and WPAN standards

The intent of IEEE Std 802.15.4 is not to compete with other wireless networking technologies but to complement the range of available wireless technologies in the lower end of the spectrum of data rates, power consumption, and cost.

Although possible for certain applications, IEEE Std 802.15.4 was not designed to overlap its application space with other wireless networking standards.

The authors strongly discourage the use of IEEE Std 802.15.4 for conventional WLAN applications. Doing so could result in a highly challenging endeavor (with very long latencies).

Table 1–1: Comparison of 802.15.4 LR-WPAN with other wireless technologies

	802.11b WLAN	Bluetooth™ WPAN	802.15.4 Low Rate WPAN
Indoors Range	~1 - 30 m	~1 - 30 m	~1 - 30 m
Data Throughput	~2-11 Mb/s	1-2 Mb/s	≤ 0.25 Mb/s
Power Consumption	medium	low	lowest
Size	Larger	Smaller	Smallest
Complexity	>6	1	0.2

LR-WPAN technology pursues applications where WPAN solutions are still too expensive, extremely low-power operation is needed and/or the performance of a technology, such as Bluetooth, is not required.

> LR-WPANs complement the range of wireless networking technologies by providing very low-power consumption capabilities at very low implementation cost while enabling applications that were previously impractical or implemented with proprietary technologies.

ISO/OSI REFERENCE MODEL

IEEE 802 communication standards define only the bottom two layers of the International Standard Organization's (ISO) Open System Interconnection (OSI) protocol reference model [29]: the physical (PHY) layer and the data link layer [19] The other layers are not specified in the standard and are normally specified by industrial consortia formed by companies interested in the manufacturing and use of the particular standard.

> The layered reference model allows the encapsulation of different levels of abstraction with a well-defined function. Each layer offers services to the layer above through service primitives.

In the case of IEEE Std 802.15.4, the Zig-Bee Alliance is an organization that has led the development of the upper layers, through the definition of application profiles. These profiles make use of a simplified five layer ISO/OSI reference model as shown in Figure 1–2. Additionally, in the recent years other wireless communication standards have used the IEEE 802.15.4 technology, as is the case of IETF 6LoWPAN and WirelessHART. The latter is the subject of Part IV of this book.

> The ZigBee Alliance is a consortium formed by several leading manufacturers, distributors, and end users focused in residential and commercial applications, including home and building automation, commercial and retail monitoring and management, and health care, among others.
>
> The HART Communications Foundation (HCF) consists of organizations supported by leading industrial process control companies in charge of managing and control of the HART standards, including WirelessHART, focused solely in industrial instrumentation.
>
> These organizations have developed the upper protocol layers including networking support and application profiles that complements IEEE Std 802.15.4 radio technology for a complete communications protocol.

Seven Layer ISO-OSI Model		Simplified Five Layer ISO-OSI Model	IEEE 802 model
7	Application	User Application	Upper Layers
6	Presentation	Application Profile ↕	
5	Session		
4	Transport		
3	Network	Network	
2	Data Link	Data Link	Logical Link Control (LLC)
			Media Access Control (MAC)
1	Physical	Physical	Physical signaling (PHY)

Figure 1–2:   ISO-OSI reference model and IEEE 802 standards model

WIRELESS SENSOR NETWORKS

Wireless sensor networks (WSNs) are a subset of wireless networking applications focused on enabling connectivity without the use of wires to sensors and actuators in general.

Sensor networks can be classified according to the type of sensors, application domain (industrial, medical, vehicular, etc.), environment where the network operates (explosive, vibration, acceleration, temperature, etc.), and by network parameters (network topology, required throughput, range, etc.). The IEEE 802.15.4 Working Group is chartered to focus on wireless sensor networks.

Because of the length of the name "Wireless Sensor and Actuator Networks," the industry has adopted "Wireless Sensor Networks" instead. In any case, it is important to remember that the design of this type of network is meant to collect and to send information to wireless transceivers attached to a sensor and/or actuator.

The interest in WSNs has three faces. First, there is a need to lower the cost of sensor installation, which comes in the form of cabling, labor, materials, testing, and verification. For example, a limit-switch might cost less than a dollar, but the installation cost could range from $50 to $100. Similarly, an industrial pressure sensor might cost few hundred dollars, but the cost of wiring it to the central control and monitoring system might reach cost as high as $300/ft. Furthermore, wiring regulations in the industrial and residential environment may require additional materials and activities for the wire installation, e.g. conduit and trained installation labor.

Second, cables require connectors that can become loose, lost, misconnected, or broken due to vibration, environmental conditions, or frequent access to neighboring devices. This problem is commonly known as "the last meter connectivity problem" and is named this way because of the analogous problem in the wide area networking called "the last mile connectivity problem."

The third aspect is that WSNs form the lower layer of intelligent maintenance systems enabling sensor-rich environments that generate abundant data that may be used to improve industrial operations. WSNs allow the gathering of more frequent data on a large number of machines and industrial systems in general. The use of large amounts of hardwired sensors networked to central systems brings a lot of complexity to the system, making it impractical in most cases.

A wireless solution for sensor networks provides flexible connectivity without the need for connectors. In addition, improved operator safety can be achieved by the use of wireless systems. This result is realized by wirelessly measuring points that a person would normally be required to interact with in a hazardous environment. Of course, WSNs share most of the issues surrounding wireless applications such as information security, authentication, small-scale radio-frequency propagation, antenna placement, and others.

Mobility is another benefit from the wireless solution, although in the WSN context, this capability is traded with "ease of installation." In other words, mobility is normally not a major requirement for a WSN system, but certain mobility concepts can be used to enable ad hoc networking. It is important to clarify that the term *mobility* in this context refers to relative motion of devices with respect to each other. For example, a wireless network within a moving machine is not considered mobile if the sensors are meant to communicate only with devices within that machine.

The set of advantages described is not enough to replace hardwired connections. The reliability and security (perceived and real) of wired networks can be higher than wireless communication systems. It is expected, however, that hybrid networks, wired and wireless, will coexist and complement each other. Wireless sensors will act as extensions of wired networks wherever the wireless capability adds value to the specific application.

Challenges in the Design of Wireless Sensor Networks

Until recently, the inertia slowing the widespread implementation of WSNs was the lack of standardized technologies that can address their requirements both at the application level and from the communications point of view. The focus of the wireless industry has been primarily on communications with higher data throughput, leaving short-range wireless connectivity behind.

An important feature required for WSNs is the capacity for easy installation of a large number of transceivers. This requirement itself has all of the issues surrounding wireless communication (e.g. range), plus other challenges unique to the application. An example of these challenges is the need to logically bind specific devices together—a lamp and its light switch.

Among the possible wireless communication technologies, we have:

- *Light-communications:* This technology includes the Infrared Data Association (IrDA®) standard, among others. The main disadvantage of this technology is the need for an unobstructed line-of-sight among the devices operating in this type of network.

- *Inductive Fields:* This technology has been extensively used for Radio Frequency Identification (RFID) applications. The main disadvantages of this technology are very low range and high energy required by the network coordinator. In addition, field alignment may be required for efficient communications.

- *Ultrasound:* Similarly to inductive fields, ultrasound requires high energy from the network coordinator. Line-of-sight is not a major concern in this technology but form factor (miniaturization) is.

- *Power Line Carrier*: This technology superimposes data signals on top of power lines. This technology has achieved enough maturity in the recent years, making it viable for a broad set of applications mainly home networking. Notice that this technology is considered wireless as no new wires are used in its operations.

- *Radio-frequency (RF)*: This technology does not require an unobstructed line-of-sight. The current state of the technology allows the implementation of low-power radio transceivers with both data-rates and ranges scalable according to the application.

RF technologies seem to provide a sufficient set of advantages for the implementation of wireless sensor networks, but they also have a set of challenges that need to address. Some of the high-level issues in the design and implementation of WSNs are described in the following section.

Power Consumption

Some applications require the use of completely untethered RF transceivers (no access to external power); this implies the use of batteries or power scavenging. If batteries are used, they should last a long time, because the need for sensor maintenance due to battery replacement goes against the original intention of ease of installation and low-cost operation. A direct implication of limited power is a limited communication range.

A common solution to overcome this obstacle is to use power cycling; that is lower the duty cycle of operation of the device. (This solution brings to surface a synchronization problem, which can be solved with the appropriate networking procedures.)

Some elementary calculations reveal that an AAA battery with capacity of 750 mAh powering off-the-shelf short-range radio transceivers (10 mA typical active current consumption) will last for more than five years if a duty cycle of less than 0.2% is maintained.

WSNs, in general, cannot make use of the "battery recharging" culture commonly associated with consumer electronic devices such as mobile phones and PDAs.

Range

Because of governmental regulations and implementation economics, RF power outputs ranging from 0 dBm to 20 dBm (1 mW to 100 mW) are typical of wireless systems operating in unlicensed bands. The limited power establishes a limited connectivity range—the maximum distance between a transmitter-receiver pair is constrained.

> The design of IEEE Std 802.15.4 allows the implementation of radio transceivers with extremely low power consumption in comparison to other wireless networking radio technologies.

In the context of WSNs, multihop network protocols are required to circumvent this problem; this in turn implies the need for suitable routing algorithms.

> An RF output power of 0 dBm with radios having a sensitivity of –70 dBm will have a range of 10 m (30 ft.) for average indoor environments (using a log-distance path loss model with path loss exponent equal to 3).
>
> Average radios today exhibit a sensitivity –90 dBm, or better providing an outdoor range in excess of 100 meters.

Availability of Frequency Bands

The RF spectrum is a scarce resource regulated by most governments. However, there are special unlicensed bands, the use of which is allowed if the wireless devices operate within a set of rules that control the RF output envelope in time, frequency, and amplitude. In some bands, in some regulation regions, regulations stipulate that in order to transmit a larger RF output power, the energy may have to be spread—i.e., spread spectrum modulation should be employed. The use of unlicensed bands is free of charge, provided that the manufacturer can demonstrate conformance with the rules.

For WSNs, these are the most common bands used (or planned to be used):

* *868.0–868.6 MHz*: Available in most European countries

* *902–928 MHz*: Available in North America

* *2.40–2.48 GHz*: Available in most countries worldwide

* *5.7–5.89 GHz*: Available in most countries worldwide

Radio devices operating in unlicensed bands must comply with local regulations. In the U.S., the Federal Communications Commission (FCC) is the institution in charge of the regulation of these devices. Similarly, the European Telecommunications Standards Institute (ETSI) coordinates regulation efforts in Europe. Other countries around the world have their own regulatory agencies; many of them accept either FCC or ETSI acceptance as proof of compliance.

IEEE Std 802.15.4 contains an informative annex focused on regulatory requirements worldwide. The material, although introductory in nature, provides the implementer with enough background information to initiate the certification endeavor.

Wireless networking standards operating in the global unlicensed bands offer potentially lower cost due to technology availability, high volumes, and limited or no re-engineering (no retuning). Recently, the major focus of short-range wireless communication original equipment manufacturers (OEMs) has been in the 2.4 GHz and 5.7 GHz bands because of the large bandwidth available.

The previous remarks identify a coexistence challenge because incompatible technologies sharing the same band will be competing to gain and maintain access. Several studies have been conducted to evaluate the impact of this issue, revealing potential problems that depend on traffic load, spatial distribution, output power, transceiver density, RF channel, and RF propagation parameters [4], [10–13], [30].

The original IEEE Std 802.15.4 was designed to operate in the following bands:
- 868.0 to 868.6 MHz 1 channel (20 kb/s, 100 kb/s, 250 kb/s)
- 902.0 to 928.0 MHz 10 channels (40 kb/s, 250 kb/s)
- 2.40 to 2.48 GHz 16 channels (250 kb/s)

The first two bands combined are called the "low band," and the third is called the "high band." This classification continues to be used even when the newest releases of IEEE Std 802.14.5™ (e.g., IEEE Std 802.15.4c/d) define other bands or use existing bands with different modulations, such as IEEE Std 802.15.4a.

Network Topology

To overcome the challenge of limited range, multihop network topologies forming a communications mesh are required. A special requirement for

WSNs is the need for low maintenance. To achieve this goal, the network topology should be designed to allow low duty cycle operation among the member devices of the network. An analysis of network topologies and network layers suitable for WSNs is presented in Chapter 6.

At this point, it is important to note an often overlooked fact: Wireless Networking implies bidirectional communications among the devices participating in the network. This characteristic in turn increases the reliability of the system and allows monitoring of the state of all the devices.

Self-Organization

For ease of installation, it is required that the wireless network be self-organizing, that is, each sensor device start participating in a network without requiring special in-situ configuration (i.e., addressing, association, and traffic balancing).

Self-organization is a feature of ad hoc networks. Ad hoc wireless networks are a collection of transceivers that form a network without the aid of any fixed infrastructure or centralized administration. The network employs routing protocols to determine the proper message path from source to destination; for example, in one protocol, each radio keeps a list of adjacent radio links and updates a routing table when the network topology changes.

The implementation of a self-organization strategy is a mixture among network topology, security implementation, and application requirements. This feature is commonly addressed by the definition of application profiles, which are not part of IEEE Std 802.15.4. Application profiles group together applications with common functionality to achieve interoperability among the devices participating in it.

Chapter 2   Low-Rate Wireless PAN Applications

...motivation for a vision

IEEE Std 802.15.4 was designed to be used in a wide variety of applications, requiring simple wireless communications over short-range distances with limited power and relaxed throughput needs. The applications that IEEE Std 802.15.4 addresses can be placed in the following classifications:

- *Stick-On Sensor*: These applications are composed of wireless sensors that can operate completely untethered, which implies battery-powered transceivers. The focus of these applications is for monitoring or remote diagnostics.

- *Virtual Wire*: These are monitoring and control applications that can only be enabled through wireless connectivity, in places where a wired communications link cannot be implemented, e.g., tire pressure monitoring, motor bearing diagnosis, and engine components.

- *Wireless Hub*: These are applications in which a centralized wireless bridge is added over a wired network. A wireless hub acts as a gateway between a wired network and a wireless LR-WPAN network. In many cases, the wireless hub network is formed with two transceivers, one being the hub and the other being an LR-WPAN device embedded in a device like a PDA.

- *Cable Replacement*: These applications attempt to add value through the removal of communication wires in wireless sensor devices. Some of these applications have been addressed by another WPAN technology: IEEE Std 802.15.1 (Bluetooth). In this case, IEEE Std 802.15.4 offers a low power alternative (at least one order of magnitude lower) with the other benefits of reduced cost. Cable replacement differs from "Stick-On" sensor applications in that the former may use a continuous source of power (mains) or rechargeable batteries.

Figure 2–1 shows the IEEE 802.15.4 LR-WPAN application scenarios just introduced.

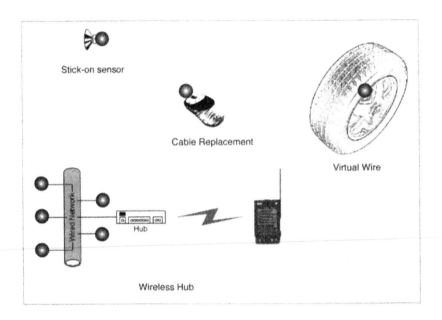

Figure 2–1: Some IEEE 802.15.4 LR-WPAN application scenarios

It is important to point out that the focus of IEEE Std 802.15.4 is to be an application enabler, where the value will be in the application and not in the wireless capability. This vision comes at odds with others, where the application is the communications capability, for example, mobile telephony.

This chapter examines some of IEEE Std 802.15.4 applications that several leading companies and visionaries have proposed and that have served as motivation for the creation of this standard.

IEEE Std 802.11, IEEE Std 802.15.1, and IEEE Std 802.15.3 were created to target specific applications. In contrast, IEEE Std 802.15.4 was designed to address a wide range of applications in different market segments always with a focus on wireless sensing communications.

INDUSTRIAL AND COMMERCIAL CONTROL AND MONITORING— WIRELESS SENSORS

As noted in Chapter 1, the interest in wireless connectivity for the industrial, commercial sector is centered on the need to lower the installation cost of sensors and actuators, while creating a sensing-rich environment as a lower layer for intelligent embedded systems.

Applications using IEEE Std 802.15.4 are mostly focused on monitoring applications with non-critical data, where longer latencies are acceptable. These industrial monitoring applications, in general, do not need high data throughput or constant updating. Instead, emphasis is placed on low power consumption to maximize the lifetime of the battery-powered devices that make up the network. Process control applications are exceptions where, although they do not require a high throughput, controlled latency and overall reliability are of prime concern. As will be shown in Part IV, IEEE Std 802.15.4 technology can provide the needed reliability for these applications.

Typical applications in industrial automation include wireless access points operating as gateways of a wired industrial protocol (e.g., HART®, Fieldbus®, PROFIBUS®, and others). These access points will allow monitoring and parameter configuration of devices connected to the wired network from a PDA (wireless hub) or will allow a wireless link to a network device (cable replacement). Figure 2–2 shows the industrial uses cases previously explained.

For the particular case of process control applications, the HART Communications Foundation is the industry group associated with the promotion and support of the IEEE 802.15.4-based WirelessHART standard also known internationally as IEC 62591.

In general, industrial applications do not have a particular network topology that characterizes them. The inherent network topology flexibility of IEEE Std 802.15.4 enables a wide range of applications in diverse industrial installations.

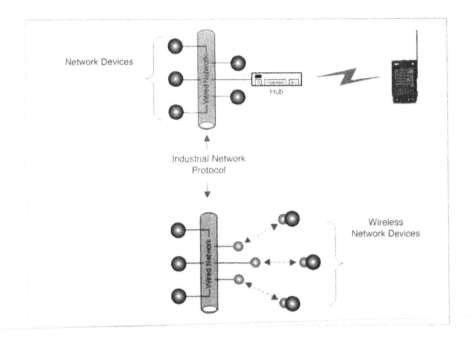

Figure 2–2: Typical industrial use cases

RESIDENTIAL SMART ENERGY

With the global energy consumption forecasted to increase two to three times by 2050, electrical energy management from generation to the consumption side has taken a prime seat in the energy sustainability discussions. The term *Smart Grid* represents a concept involving the digital upgrade of the electrical transmission and distributions grid with the purpose of performing peak consumption management, integration of distributed alternative energy production, and enable consumers to have greater control over energy usage.

Initially, peak management is envisioned to be triggered by the electrical utilities using a process called demand response, or remote enablement of major customer appliances such as the HVAC system, plug-in hybrid charger, etc. Another peak management technique is based on dynamic pricing where the utility sends price signals to the customer facilities to entice them to shift the use of large electrical loads to times where price is lower (presumably at moments where the system contains extra capacity to handle such loads).

In the past, a major challenge to perform the Smart Grid vision at the customer level has been related to the residential market. Houses in general lack a standard communications infrastructure to allow the proper coordination and management of loads for both demand response and dynamic pricing. However, this issue is mitigated with the use of 802.15.4-based networks. Low-cost home area networks (HANs) enabled by 802.15.4 systems allow easy retrofitting in the residential space while allowing appliance-level monitoring and control functions required for energy management (remote by the utility or locally by the customer).

Recently, ZigBee's 802.15.4-based protocol created an application profile to enable the creation of a special type of Home Area network that defines energy monitoring and control functionality. This application profile was named Smart Energy.

In 2009, the U.S. Government selected ZigBee's Smart Energy Profile as the wireless communications standard for home area networks interfaced with the Smart Grid.

HOME AUTOMATION AND NETWORKING

The consumer and home automation market presents a significant potential because of its size. IEEE 802.15.4 LR-WPAN devices replace wires in the home environment at a very low cost. Types of potential wireless sensing devices include consumer electronic equipment, personal computer peripherals, interactive toys and games, as well as home security, lighting control, and air conditioning systems [2]. Most of these applications have an industry group interested in using a low-cost wireless solution. Some of these industry groups are as follows: the Consumer Electronics Association (CEA), HomePlug Alliance, and the ZigBee Alliance. Of particular interest is the analysis performed by the ZigBee Alliance that identifies home automation applications as providing the lower bound on required data throughput and the upper bound on acceptable message latency for LR-WPANs.

IEEE Std 802.15.4 adopted a maximum data rate of 250 kb/s as a result of an analysis performed for home automation applications.

Consumer Electronics

The consumer electronics sector includes radios, televisions, VCRs, CD players, DVD players, remote controls, and appliances in general. The opportunities to bundle products and services with a common control mechanism (IEEE Std 802.15.4-enabled remote) is providing many interesting opportunities. With the convergence of PC and consumer electronics markets, IEEE Std 802.15.4 gives the end user the ability to control devices from both segments with the same intuitive device they are familiar with: the remote control. It will now be possible to have remote control units assigned to individuals instead of being assigned to a piece of equipment (in addition to reducing the many remote control units in the home). Individualization and two-way remote controls become possible, to enable feature rich programming and control. In addition, communication links between multimedia devices to set volume levels, equalizer, and other control settings within a home entertainment center, and home systems such as thermostat and security functions control will soon be possible under a common communications infrastructure enabled by IEEE Std 802.15.4-based systems.

In June 2008, the RF4CE Consortium was formed with the purpose of creating a remote control standard for consumer electronics based on IEEE Std 802.15.4. The aim of this organization is to replace infrared remote control technology over five years, providing a faster and more reliable way to operate consumer electronic devices

In March 2009 the RF4CE Consortium decided to merge with the ZigBee Alliance to deliver a standardized specification for RF-controlled home entertainment products that complements the ZigBee vision.

without needing line-of-sight. Furthermore, the two-way communications capability provided by IEEE Std 802.15.4 enables advanced features between the device and the remote control, creating an enhanced customer experience.

PC Peripherals

PC peripherals, including wireless mice, keyboards, joysticks, PDAs, and games, represent a large segment inside the home for low-rate wireless connectivity. In addition, PC manufacturers and software companies are trying to change the interface to the PC, to transform it into a more consumer-oriented device. With the advent of remote controls and PDAs, this will become the way a PC is controlled. IEEE Std 802.15.4 technology is uniquely suited to provide the kind of low-power, low-cost solutions this application scenario needs.

Home Automation

The home automation sector, which includes heating, ventilation, and air conditioning (HVAC), security, lighting, and the control of objects such as curtains, windows, doors, and locks, represents certainly a great opportunity for wireless innovation inside the home. In many cases, the thermostats in houses are located in the rooms where minimal time is spent (corridors, foyers, etc.). As a result, the temperature readings often do not correlate well with the temperature experienced by those living in the home, and hence, control of the environment in the home is often inefficient.

Wireless thermostats can be dispersed throughout the home, and they can be linked with wireless vent controllers (zoning controllers) in the HVAC system to regulate the temperature in rooms individually. Window curtains can be drawn when the television that turned on during daylight hours; wall clocks can be coordinated with a master reference so they are reset after power failures automatically (no more blinking *12:00* on the microwave!). Home security can be as simple as the automotive remote keyless entry key fob—a single button press could lock all doors and windows; wireless smoke detectors and glass breakage sensors will connect to the home's security system. With the availability of all of these applications using IEEE Std 802.15.4, a common remote control can be used to control all of these functions in addition to the consumer electronics devices.

Some lighting manufacturers have announced recently the integration of IEEE Std 802.15.4 transceivers in light bulbs and ballasts. This practice would eliminate wiring (especially important for retrofit) so that standard wireless switches can be placed anywhere in the home without the need to break walls.

In November 2009, the ClimateTalk Alliance was launched. This organization is chartered with defining a common information model that leverages ZigBee's Smart Energy Profile while defining a standardized HVAC messaging system.

Home Security

Similarly to home automation, security sensors are dispersed around the home environment. Typical security sensors consist of motion sensors, door, and window open/close sensors, water leakage, and other specialty devices. Again, the value proposition of a wireless capability using LR-WPAN transceivers is based on the ease of installation because their wired sensor counterpart requires hiding wires and drilling through walls and floors.

Wireless sensors in home security face two important challenges. The first one is related to power consumption of the sensor. Commonly, these sensors are not mains-powered, because they are usually located on windows and doors. IEEE Std 802.15.4 overcomes this challenge, which, as stated before, was designed with a simple yet efficient protocol stack. Furthermore, the medium access control (MAC) definition offers the upper layers an interface with full control to enable or disable the radio functionality. The second challenge is related to wireless range due to the propagation of radio frequency signals. The output power of IEEE Std 802.15.4 transceivers is bounded due to governmental regulations and by the need to conserve power (when necessary). IEEE Std 802.15.4 offers peer-to-peer protocol support that allows a network layer (not included in the standard) to implement a multihop network, thereby increasing the effective range.

Personal Healthcare

The personal health care sector includes sensors, monitors, and diagnostics. This field is separate from medical telemetry; personal health care includes items such as pedometers and pulse rate monitors used by joggers and other athletes. These devices produce data that often need to be transferred to a display device; a wireless connection is often the only practical method to do so. Another application is health record maintenance; a person's daily weight, temperature, and so on can be easily transmitted by a wireless link between the proper measuring device (scale, thermometer, etc.) and recorded on a personal computer or a PDA. With the availability of this information on the Internet, health care professionals can access this data and monitor the status of persons remotely—no more need to drive to the weight loss center every week.

In the recent years, 802.15.4 technology has been the key enabler for elder-care applications. This application consists on remote monitoring of the daily activities of seniors living independently to estimate health status. This application allows elder home owners to stay in the comfort of their own homes while having peace of mind that in the event of a health event, health care providers are alerted while family members keep tabs on their loved ones from afar. Among the multiple sensors used for this application, the most publicized is the wireless *panic button* that seniors use to send an emergency signal if they suddenly become incapacitated.

> The ZigBee Alliance is defining a Health Care application profile with the goal of enabling wireless communication between devices and services that foster safe, healthy, and independent living conditions for the disabled or elderly.

Toys and Games

Low-rate wireless networks have much to offer in the toys and games sector, especially when the communication is among toys or a toy and a personal computer. Toys are extremely cost-sensitive; the limiting factor on such computationally intensive tasks as voice recognition and synthesis in a toy often is not the technical feasibility of the required task, but the need to perform it for a low total hardware cost. By adding a wireless link between the toy and a nearby personal computer, the cost of the toy can be reduced, because the toy only needs to include the wireless link and the necessary sensors and actuators (e.g., microphones and speakers). Such PC-enhanced toys exhibit sophisticated behaviors (e.g., mobile robots), limited only by the capabilities of the computer and the wireless link. Other opportunities in this sector include wireless gaming between individuals or groups; in addition to simple wireless links between players, both toys and games could receive updates when within range of a computer and then operate autonomously in the new mode, or with the new personality or feature, when leaving the range of the computer. Wireless gaming manufacturers using IEEE Std 802.15.4-based solutions can address the installed base of games in addition to the new games, because the cost and power consumption of a wireless add-on module is targeted to be low enough to make it economically enticing to users to add on to their existing games.

A summary of the performance requirements of the home automation and networking application is shown in Table 2–1. Note that the typical data rate for most applications within each sector is significantly below the maximum rate shown.

Table 2–1: Requirements of home networking market segments

Sector	Max Required Data Rate	Maximum Acceptable Message Latency
Consumer Electronics	3 kb/s	16.7 ms
PC Peripherals	115.2 kb/s	16.7 ms
Home Automation	10 kb/s	100 ms
Personal Healthcare	10 kb/s	30 ms
Toys and Games	115.2 kb/s	16.7 - 100 ms

AUTOMOTIVE SENSING

Wireless communication is finding its way into cars, as driver comfort and the number of features offered increase. The first wave of wireless technology in a car came from the remote keyless entry application and its derivatives. The second wave is happening now in the form of "cable replacement" for telematics applications. The IEEE Std 802.15.1/Bluetooth technology has taken the lead in addressing this type of application, maintaining a major focus on telephony applications such as handling of a mobile phone from embedded devices in a car (in-car hands-free speech) and car personalization through data exchanged from a mobile phone to the car.

In general, cost plays a fundamental role for vehicle applications not related to safety and luxury convenience. Today, a car is loaded with a great amount of sensors and actuators distributed all around the vehicle. The use of these devices has originated a significant growth in wiring, causing a great impact on the installation costs, diagnostics, maintenance, and even fuel consumption!

The wireless option introduces flexibility in installation and an advanced alternative to wired connections. A special challenge for automotive applications is meeting the harsh automotive environment while maintaining the low-cost requirements.

An application example in the category of virtual wire is a tire pressure monitoring system. The system consists of four pressure sensors, mounted on each tire, and a central station to receive the collected data. Because the pressure sensors have to be mounted onto the tires, this application does not permit the use of any communication wires or power cables. Therefore, sensors have to be battery powered. Because it is impractical to replace the sensors or their batteries between tire changes, the sensor batteries are required to last at least three years and preferably five years. This requirement puts significant constraints on the power consumption of the electronic components and requires power management capabilities. The data to be communicated (the measured tire pressure) are, in most cases, only a few bits in size. This information is transmitted about every 1 to 10 minutes under non-alarm conditions. Unless there is a fast loss of pressure, the message latency is not of significant concern. In case of a sudden pressure loss, the central control unit is notified immediately, in which case, the power consumption is not of concern because most likely the tire has to be replaced. Extreme automotive environmental conditions and the metallic structure of the car complicate the RF design. However, the use of two-way communications helps significantly in augmenting the reliability of the communications.

PRECISION AGRICULTURE

Another application for LR-WPANs is precision agriculture, also known as precision farming. Precision agriculture is an environment-friendly system solution that optimizes product quality and quantity while minimizing cost, human intervention, and the variation caused by the unpredictability of nature. Today, agriculture still demands much of both user and environment. It is mainly hardware-oriented with manual and on-site control using independent dumb machines, which produce unpredictable quality and quantity. With the new paradigm of precision agriculture, farming would become more information and software-oriented, using automatic and remote-controlled, networked smart machines. This application requires large mesh-type networks consisting of potentially thousands of LR-WPAN devices linked with sensors. These sensors will gather field information such as soil moisture content, nitrogen concentration, and pH level. Weather sensors for measuring rainfall, temperature, humidity, and barometric pressure also provide the farmer with valuable information. Each sensor passes the measured data to its corresponding LR-WPAN device, which in turn will pass it through the network to a central collection device. For the sensor data to be useful, location-aware technology is necessary to correlate each sensor with its specific location in the field. The combined information gives farmers an early alert of potential problems so that they can achieve higher crop yields. The precision agriculture application is at the low end of the LR-WPAN application range, requiring the transmission of only a few bits of data per day by each deployed device. The data flow will be asynchronous in nature, with minimal restrictions on data latency. This combination of factors is advantageous for achieving long battery life. This application requires the use of a mesh topology where some devices serve as repeaters for others, relaying messages to the final destination. This application requires the network to be self-configuring because manual setup of a network of the proposed size is not feasible.

MISCELLANEOUS APPLICATIONS

A unique application scenario, falling within the consumer market, is a classroom calculator network. The teacher's workstation or the PAN coordinator would send tasks and math problems to each of the student's graphic calculators or the network devices. After completion, the students would upload their solutions back to the teacher's workstation. This network would need to support only a small number of devices, and it would require disallowing any peer-to-peer communication to prevent students from exchanging the solutions. The typical payload would be 100 to 500 bytes of information, sent several times per student per hour. It is desired that the batteries powering the calculator and communication function last the duration of a semester.

Another important application in the category of wireless hub is remote metering and remote configuration. This type of application enables improved hazard protection and operator convenience. Devices using an inexpensive IEEE 802.15.4-based LR-WPAN device can exchange information wirelessly using PDAs, which can have previously programmed configuration information or contain software that synchronizes the data with large databases.

Part II

Chapter 3   IEEE Std 802.15.4 Technical Overview

...warming up

The IEEE Std 802.15.4 Working Group established the goal of ultra-low complexity, low-cost, and extremely low-power wireless connectivity among inexpensive fixed, portable, and moving devices. With relaxed throughput and latency requirements, low-cost and low-power design could be achieved. The shared dream of ubiquitous, untethered short-range communications began to be realized.

Substantial costs can be associated with administration, sales, and marketing activities, as well as the cost of manufacturing and operation of wireless products. To achieve the goal of low total product cost as well as long battery life for low operating cost, IEEE Std 802.15.4 provides reasonable trade-offs in several performance metrics. To control the administrative costs of both implementer and user, IEEE Std 802.15.4 devices employ unlicensed radio bands. Because IEEE Std 802.15.4 service is designed to be a short-range service without infrastructure, and able to support very large networks, both transmit and receive functions can and must consume little power and have a low operating cost. This chapter presents some of the unique features of IEEE Std 802.15.4 that enable this performance vision to be met.

HOW LOW-COST AND LONG BATTERY LIFE ARE ACHIEVED

Duty Cycle

The battery is a relatively high-cost component in a transceiver system. The battery provides the energy for the communications, but at a cost of operation and replacement. Total system cost could not be minimized if a high-cost, exotic battery technology was required. The characteristics of low-cost batteries limit the choices for other attributes of the system. For example, the fully charged energy capacity and the instantaneous power delivery capacity of most batteries are interrelated. The fully charged energy capacity is achieved only if the instantaneous power delivery requirement is limited. Because of this limitation, to achieve long battery life, the energy must be taken continuously at an extremely low-rate, or in small amounts at a low-duty cycle. Because of high power consumption of practical wireless circuits, it is impractical to attain the desired battery life under constant operation; IEEE Std 802.15.4 was therefore designed to support very low duty-cycle operation. The standard allows some devices to operate at low-duty cycles with both the transmitter and receiver inactive for over 99% of the time they are in operation.

In practical implementations there will always be a small amount of standby power consumed for timers, and so forth, when the wireless circuits are inactive. To reduce the time-average power consumption, both the active power and the standby power of an implementation should be reduced as much as possible. However, for a given technology and set of applications supported by the network, there is a practical limitation for both the active power and the standby power. Furthermore, for most applications, the active power is much larger than the standby power. Under this condition, we can see that by reducing the duty cycle low-power consumption levels, and associated long battery lifetimes can be reached.

For example, for a device with 10 mW active power and 10 μW standby power, if the duty cycle is 0.1%, then the time-average power drain is about 19.99 μW. If the device is supplied by a 750 mAh AAA battery, linearly regulated to 1 V, it will have a battery life of more than 37,000 hours, or more than four years.

To support low-duty cycles, the IEEE Std 802.15.4 beacon packet can be as short as 544 μs in the 2.4 GHz band, while the superframe period (the time between network beacons) may be extended from 15.36 ms to over four minutes. The result is a beacon duty cycle that may be set from 2.3% to 0.000216% (2.16 ppm). Further, the standard supports a non-beacon mode in which a network may operate without any beacons. The non-beacon mode enables the slaves in a master-slave star network, for example, to remain in standby mode indefinitely, only contacting the master (which may be mains powered, perhaps, and therefore capable of constant reception) when an event occurs. The slaves can therefore have an almost unlimited battery life, limited primarily by their standby power consumption. The non-beacon mode, in fact, helps meet regulatory requirements for operation in the 868 MHz band, which has a maximum duty cycle limitation of 1%.

Modulation

With its low-cost and low-power consumption goals in mind, the IEEE Std 802.15.4 communications protocol was designed to support digital data communication only (i.e., analog service is not supported). Data-only service allows the modulation scheme to be chosen to be highly efficient, enabling a low-cost implementation. Likewise, the protocol supports only half-duplex operation so that the transmitter and receiver are not required to be active simultaneously.

The physical layer employ a spreading sequence to provide the benefits of direct sequence spread spectrum (DSSS) service. For example, the 2.4 GHz PHY employs a form of multilevel orthogonal signaling, sending 4 bits per symbol, which simultaneously enables both a high-data rate (to return to standby mode quickly) and a relatively low-symbol rate (to minimize active power). In the 868/915 MHz PHY the chip modulation schemes are either raised-cosine-shaped Binary Phase Shift-Keying (BPSK), half-sine-shaped Offset Quadrature Phase Shift-Keying (O-QPSK), or root-raised-cosine-shaped Parallel Sequence Spread Spectrum (PSSS). The 2.4 GHz PHY uses a half-sine-shaped Offset Quadrature Phase Shift-Keying (O-QPSK) chip modulation scheme. Further, the half-sine-shaped O-QPSK maintains a peak-to-average carrier power ratio of one, which minimizes both power consumption and implementation complexity.

Use of Direct Sequence Spread Spectrum

DSSS is one of several techniques to increase the bandwidth of a transmitted signal. The wideband techniques provide improved communications qualities but usually sacrifice spectrum utilization. For instance, the conventional broadcast FM service employs a form of spectrum spreading (wideband FM) that, although it does not employ DSSS, exemplifies the improved signal-to-noise ratio (SNR) communications benefits possible with wideband techniques.

DSSS can be modeled as applying a prearranged pseudo-random digital sequence to directly phase-modulate the already data-modulated carrier, at a rate in excess of the data rate. The resulting DSSS signal occupies a much greater bandwidth, albeit with a lower spectral power density. The signal is recovered by demodulating the received signal with a replica of the same modulating pseudo-random digital sequence. The replica is generated in the receiver by a technique that ensures that the replica is closely coherent with the modulation in the transmitted signal. An ideally recreated replica transforms the received DSSS signal back into a replica of the transmitter's original data-modulated carrier. The DSSS process spreads the original signal into a wider bandwidth for transmission over the channel, and then despreads the signal at the receiver to recover the original signal and the information it contained.

The benefits of the DSSS form of spectrum spreading are a direct result of the relative coherence of the pseudo-random digital sequences used to phase modulate and recover the replica of the original data-modulated carrier. Only those signals that are modulated by the correspondent coherent pseudo-random digital sequence are despread back into a replica of the original data-modulated carrier. All other signals are further spread by the receiver despreading phase-modulation process.

Because a spreading process reduces the power spectral density of noncoherent signals being modulated, interfering signals, adjacent channel signals, and even other noncoherent spread-spectrum devices on the same channel have a lowered spectral density at the receiver within the narrow bandwidth of the original data-modulated carrier. All other signals are spread to a wider bandwidth, lower spectral density, and smaller energy contribution when filtered to the original data-modulated carrier bandwidth. This processing gain can be used, for example, to reduce requirements on channel filters, thereby lowering implementation costs.

The cost benefits of DSSS result from an SNR improvement of coherent versus noncoherent signals, but there are other pertinent effects also.

One effect is that the power spectral density of the signals emitted from the spread spectrum transmitter are lowered relative to the power spectral density of the unspread data-modulated carrier, and are less likely to interfere with narrowband services using the same band. This coexistence effect enables the use of multiple services in the same bandwidth. A DSSS service can be added to existing spectrum allocation with reduced interference to existing services and likewise be relatively unaffected also by those services. DSSS is a nearly ideal means to share spectrum.

A more subtle advantage of DSSS is that its implementations may be made of largely digital circuits, with few analog circuits. Because digital circuits follow Moore's Law and shrink with improving integrated circuit lithography, they and the resulting transceiver implementation become cheaper to produce over time.

Finally, because of these favorable characteristics of spread spectrum operation, use of DSSS (or another form of spread spectrum, frequency hopping) is mandated by many regulatory bodies worldwide for the unlicensed bands under their jurisdiction. IEEE Std 802.15.4 makes use of these spectrum allocations to define a practical low-power, low-cost, unlicensed service.

> Moore's Law establishes that the number of transistors on an integrated circuit doubles every two years.

Transmitter Power

Additionally, although IEEE Std 802.15.4 permits any legally acceptable output power, it requires only that the compliant device be capable of transmitting −3 dBm, well within the instantaneous power capacity of inexpensive battery sources, and within the capability of highly integrated and low-cost system-on-a-chip (SoC) implementations.

Typical radio implementations of IEEE Std 802.15.4 offer on-chip output power ranging from 0 dBm to 4 dBm. Higher output power will require the use of an external power amplifier circuitry (which impacts overall system cost and power consumption).

> Many IEEE Std 802.15.4 semiconductor manufacturers design their chips to offer a variety of output power ranges ranging from −10 dBm to +10 dBm.

Receiver Sensitivity

The required minimum −92 dBm RF sensitivity specification in the 868/915 MHz PHY, and the −85 dBm RF sensitivity specification in the 2.4 GHz PHY, permit the use of a simple receiver to achieve compliance. Simple, low-cost receiver designs, with little radio-frequency amplification (which can be expensive in terms of power consumption) are possible.

Thanks to the differentiating efforts from multiple IEEE Std 802.15.4 chip suppliers, the minimum sensitivity levels mandated by the standard has been improved in almost all practical chip implementations.

Quality of Service

To reduce implementation complexity, IEEE Std 802.15.4 does not support isochronous communication, and it does not support multiple classes of service within a single PAN. An exception is that the standard does support the optional use of Guaranteed Time Slots (GTS), which reserve network time for synchronous communication (avoiding potential channel access delays), at the discretion of the contention-free period (CFP). This extends the application space of IEEE Std 802.15.4 devices to include applications requiring low-latency communications, such as wireless joysticks and mice, or more demanding applications such as industrial process controls.

Although IEEE Std 802.15.4 is of unusually low complexity, the PHY layers support instantaneous link data rates of 20 kb/s, 40 kb/s, 100 kb/s, and 250 kb/s. The data rates are high relative to the data throughput in order to minimize device duty cycle.

In summary, IEEE Std 802.15.4 is designed to trade lower data throughput and higher message latencies for significantly lower cost and power consumption, while still providing useful communications for a wide class of services.

OTHER NOTABLE IEEE STD 802.15.4 FEATURES

Network Components

An IEEE Std 802.15.4 network is composed of a set of IEEE Std 802.15.4 wireless devices. Each network contains exactly one specialized central network coordinator, called the PAN coordinator. Only the PAN coordinator can establish a new network, and it defines the structure and operating mode of the network. Other devices can join the network by requesting permission from the PAN coordinator. Besides the PAN coordinator, the standard distinguishes between two additional logical device types, one being a coordinator and the other, a network device. A coordinator is a device that can provide coordination services to other devices of the network, such as acting as a proxy for network devices that are not within range of the PAN coordinator. An 802.15.4 network consists of one PAN coordinator and at least one network device.

IEEE Std 802.15.4 defines two types of devices, the full function device (FFD) and the reduced function device (RFD). The FFD contains the complete set of MAC services and allows it to operate as any of the three possible personalities (PAN Coordinator, Coordinator, or Network Device). The RFD contains a reduced set of the MAC services, and it can only operate as a network device.

RFDs were defined to allow the implementation of extremely simple devices that require minimal resources in terms of processing power and memory capacity. These characteristics have a direct impact on the implementation of lower cost devices to enable applications such as light switches, stick-on sensors, and others.

Multiple Network Topologies

An IEEE Std 802.15.4 network may operate in one of two basic topologies. The first topology is called a star topology, and it is formed around a full-function device that is designated the PAN coordinator and that acts as a hub with a collection of additional full or reduced function devices that act as data terminal locations. In this regard, the PAN coordinator performs a special function in the network; it is the only device in the network that forms a direct link to more than one other device.

The second topology enables peer-to-peer communication without the direct involvement of a designated network coordinator as such (although a PAN coordinator is required somewhere in the network), and each device is able to form multiple direct links to other devices. The capability to form direct links with more than one other correspondent device ensures that the network graph can form redundant paths, but the complexity is increased. As a consequence of this second topology, many types of peer-to-peer network architectures can be supported.

One such complex network that can be supported by IEEE Std 802.15.4, but is not part of the specification, is the cluster-tree. A cluster-tree (see Figure 3–1) can be interpreted as a hierarchical tree of network device clusters. The cluster-tree arrangement provides complete connectivity, somewhat simplified routing, and a minimum number of direct links compared to a fully interconnected peer-to-peer structure, but it may result in increased data traffic latency.

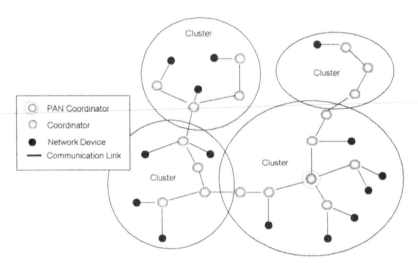

Figure 3–1: Cluster-tree network

The leaves of a cluster-tree (those devices at the edge of the hierarchy) may be either FFDs or RFDs, since they do not need to relay messages for other devices. However, all other devices in the network must be FFDs. Exactly one device in the network assumes the special role of the PAN coordinator, which has a function much like the root device of a hierarchical tree. The PAN coordinator may be selected because it has special computation capability, a bridging capability to other network protocols, or simply because it was the first participant in the formation of the network. Other devices may function as roots of individual clusters; these are designated as cluster heads, or coordinators. The clusters form a hierarchical tree with typical parent-child relationship between devices. In this cluster-tree network example, all leaf nodes (devices at the end of the branches) are considered network devices, while all other devices need to be coordinators.

Network topologies built on the IEEE Std 802.15.4 peer-to-peer communication feature are supported by the higher levels of the protocol stack which are outside the scope of IEEE Std 802.15.4. They are introduced here to show the flexibility of which the standard is capable; additional discussion is in Chapter 6.

Channel Access

Regardless of the type of network employed, each network device employs a carrier sense multiple access-collision avoidance (CSMA-CA) protocol to avoid wasteful collisions when multiple simultaneous transmissions might otherwise occur. Exceptions to this are beacon transmissions, transmissions in GTSs, and acknowledgments, each of which is transmitted without the CSMA-CA protocol.

The CSMA-CA protocol is based on the shared nature of the RF channel. Whenever two or more senders are active on a channel simultaneously, the probability that any one of them is successful in sending a message decreases due to collisions and their mutual interference. In the RF environment, a great deal of the actual interference depends on the location of the competing transmitters and their respective correspondents, but the location information is often not available to the transmitting devices. One way to avoid collisions in the channel is to listen first and to transmit only if the channel is clear. The carrier sense improves the probability of access to the clear channel and the possibility of a collision is reduced. In this fashion, the channel capacity is utilized more efficiently.

Multiple PHYs

IEEE Std 802.15.4 along with all its amendments defines a total of 10 PHYs—one mandatory and two optional PHYs for the 868 and 915 MHz bands, one mandatory for the 2.4 GHz band, three optional ultra-wide band PHYs, one optional Chirp spread spectrum band for the 2.4 GHz band, two optional PHYs for the 780 MHz Chinese band, and two optional PHYs for the 950 MHz Japanese band.

Note that the 868/915 MHz PHY requires that a compliant device be capable of operating in both the 868 MHz band and the 915 MHz band. This stipulation reflects the desire to minimize the number of potentially incompatible products on the market, and it recognizes that there is little cost penalty in this requirement, due to the small frequency difference between the two bands.

The Task Group that developed IEEE Std 802.15.4 was comprised of volunteers from around the world and therefore considered that compliant devices may be developed and used worldwide. To keep administrative costs low, IEEE Std 802.15.4 assumes that all devices will perform in unlicensed bands that each would fall under the provisions of the regulatory agency with jurisdiction for the particular service area. Unfortunately, the regulatory agencies worldwide have allocated the spectrum differently between and among themselves. There are three bands available for this type of service—one worldwide and four regional special cases.

For implementers, the choice of bands depends on more than simple technical considerations. The 868 and 915 MHz bands, although they are likely to be less crowded and may offer better QoS, are not available worldwide. The decision to implement products for a limited market, albeit with possible better service, falls in the realm of marketing and business rather than purely technical matters. There are hidden costs of distribution and marketing of otherwise identical products that differ only in frequency band. There are interference problems that bring up questions of how to keep products from moving from one regulatory region to another. Other applications (like wireless luggage tags) do not fit into a regional marketing strategy and do not present a consideration, but the possibility of regional strategies exists for some products.

Portions of the 2.4 GHz band are available nearly worldwide; the 868 MHz band is available in Europe, and portions of the 915 MHz band are available in North America, Australia, New Zealand, and some parts of South America. IEEE Std 802.15.4 defines a single channel with a mandatory raw data rate of 20 kb/s and two optional data rates of 100 kb/s and 250 kb/s in the 868 MHz band. In the 902—928 MHz band, the standard defines 10 channels with a mandatory raw data rate of 40 kb/s (doubled chip rate with respect to the 868 MHz band) and an optional raw data rate of 250 kb/s. Higher in the spectrum, at 2.4 GHz, IEEE Std 802.15.4 defines 16 channels with a raw data rate of 250 kb/s.

The amendment 802.15.4a defines three ultra wide bands at frequencies of 3 GHz to 5 GHz, 6 GHz to 10 GHz, and less than1 GHz offering a mandatory data rate of 851 kb/s with options to 110 kb/s, 6.81 Mb/s, and 27.24 Mb/s. In addition, this amendment defines a chirp spread spectrum (CSS) PHY that operates at the 2.4 GHz band with a data rate of 1000 kb/s with an option to 250 kb/s.

The amendment 802.15.4c defines a new PHY designed uniquely for the Chinese market operating at the 779–787 MHz band. This band was designated by the Chinese government for wireless sensor applications and provides a data rate of 250 kb/s. Similarly the amendment 802.15.4d defines a new PHY developed for the Japanese market operating the 950–956 MHz band offering 100 kb/s with an option to 20 kb/s.

Error Control

IEEE Std 802.15.4 employs a simple full-handshake protocol to ensure reliable data transfer and good QoS. With the exception of broadcast frames (e.g., beacons) and the acknowledgment frame, each received frame may be acknowledged to assure the transmitting device that its message was, in fact, received. If a requested acknowledgment frame is not received by the transmitting device, the entire transmitted frame may be repeated.

To detect that a message has been received correctly, a cyclic redundancy check (CRC) is used. The message bits are treated as a long binary number and divided by a relatively large prime number. The quotient of the division is discarded, and the remainder of the division is transmitted along with the message. The same division with the same prime number is performed at the receiver, and a match signifies a high probability of an uncorrupted communication.

FOUR FRAME TYPES

There are four frame structures, each designated as a PHY Service Data Unit (PSDU) in the standard for data transactions—a beacon frame, a data frame, an acknowledgment frame, and a MAC command frame. As shown in Figure 3–2, all frames are structured in a very similar fashion, with the primary differences in their purpose or payloads. Each PHY Protocol Data Unit (PPDU) is constructed with a Synchronization Header (SHR), a PHY Header (PHR), and a PHY Service Data Unit, composed of the MAC Protocol Data Unit (MPDU) as a data structure that services the MAC protocol layer. The MPDU is constructed with a MAC Header (MHR), a MAC Footer (MFR), and a MAC Service Data Unit (MSDU)—excepting the MPDU of the acknowledgment frame, which does not contain an MSDU.

The MSDU is a data (i.e., payload) field component of a given frame containing information pertinent to the MAC services supported by the frame, including superframe identification and sequencing information, addressing information, and other information.

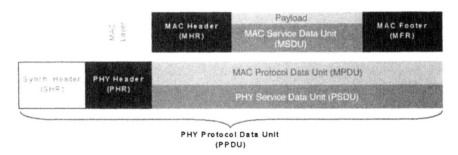

Figure 3–2: Frame structures format

Beacon Frames

Transmission of the beacon frame is available only to FFDs in the network, regardless of whether the network is a star, cluster-tree, or other topology. The beacon frame is provided as a service originating in the MAC protocol layer and interfaced to the PHY protocol layer. It has a number of uses, including superframe boundary marker, frame synchronizing signal, and association supervision, all as a service to the higher protocol layers.

As a superframe boundary marker, the beacon frame provides a timing reference to designate the boundaries and structure of a superframe. In the higher level of the protocol, the superframe permits a fixed number of frames to be designated between beacon frames, whether each frame position is occupied or not.

As a frame synchronization signal, the beacon frame allows the synchronization of the superframe to a known starting time, preventing collisions, allowing receivers to sleep whenever there are no transactions, and permitting low message latency by limiting the typical delay of traffic to the length of one superframe on any one link. Further, the synchronization allows the network device transmitting the beacon to become the timing reference for link communication and allows a relaxation of an individual network device's timing requirements.

Each network device may have a less precise timing reference that is periodically adjusted to match the beacon device's reference.

Data Frames

The data frame is available to all devices in any network, regardless of whether the network is a star, cluster-tree, or other topology. The data frame, like the beacon frame, is provided as a service in the MAC layer due to a request from a higher layer. It is interfaced to the PHY layer. It provides the primary data payload as a service to the higher protocol layers.

Acknowledgment Frames

The acknowledgment frame is also available to all devices in any network, regardless of whether the network is a star, cluster-tree, or other topology. The acknowledgment likewise is provided as a service originating in the MAC protocol layer and interfaced to the PHY protocol layer. It provides only the acknowledgment for receipt of data as a service to the higher-level protocol layers for end-to-end message control. The acknowledgment frame does not contain a MAC payload.

MAC Command Frame

The MAC command frame is also available to all devices in any network, regardless of whether the network is a star, cluster-tree, or other topology. The MAC command frame, like the other frames, is provided as a service originating in the MAC protocol layer and interfaced to the PHY protocol layer. It provides the primary supervisory payload as a service to the MAC protocol layer.

For each of these four frame types, the MPDU is further encapsulated with an additional preamble and PHY header to form the physical layer protocol data unit (PPDU). This complete data structure allows a receiver to achieve symbol synchronization with the frame and obtain the frame length information.

Chapter 4 Physical Layer

...from bytes to watts

The physical layer (PHY) provides the interface with the physical medium where the actual communications occurs. The PHY layer is the lowest component in the ISO/OSI model and is in charge of providing control (activation and deactivation) of the radio transceiver, energy detection, link quality, clear channel assessment, channel selection, and the transmission and reception of message packets through the physical medium [29].

Governments around the world regulate and administer the radio-frequency spectrum. This spectrum is divided in bands allocated for multiple services and applications covering telephony, networking, military communications, TV, radio, etc. In most cases, each of these bands require a special license for operation in such bands. However, special types of bands are reserved for unlicensed operation. These bands can be used without the need of any special licensing as long as they are used within constraints established by the local authorities such as limits in output power, duty cycle, modulation, and other parameters.

The particulars of unlicensed services vary throughout the world, but generally, the devices must conform to a set of regulatory limitations that are specific for each application. Transmitters for each operation must pass a testing protocol to ensure that they meet the specific requirements. After a particular design has passed the regimen of testing, it is issued a certification or type approval for that service or product. IEEE Std 802.15.4 devices are targeted toward type approval.

Implementations of IEEE Std 802.15.4 must conform to local regulations of the country where it will operate. In the case of Europe, Japan, Canada, and the United States, the regulation consists of unlicensed but type-approved DSSS services. It is the constraint for unlicensed but type-approved compliance that dictates the frequency bands of operation, as well as a few of the other characteristics of the service.

IEEE Std 802.15.4 is written so that conforming devices can be manufactured to operate in any of three particular bands. Two of these bands are limited to specific geographical regions, but one band is available for nearly worldwide service. Details of these bands are presented next.

In Europe, the European Telecommunications Standards Institute (ETSI) publishes recommendations that are recognized by all European regulatory agencies, but each service area falls under individual national type-approval authority. Within the European service area, one common band operates between 868.0 Mhz and 868.6z MHz that supports a single channel of low data rate service with less than 1% transmission duty cycle [5], [6].

In China, the Chinese government recently created regulations that allow the creation of a unlicensed wireless sensor network spectrum operating in the 779–787 MHz band. Similarly, the Japanese government also approved the creation of a unlicensed WSN band but with operation in the 850–856 MHz band.

Within the United States, the national regulatory agency is the Federal Communications Commission (FCC). The FCC has specific authority only within the United States; however, the FCC regulations are used as a model by many other nations in the Americas and the Pacific Rim. Just as in Europe, a unique band is available for service; this band is called the 915 MHz band, and it covers the range between 902 MHz and 928 MHz. This band enables the provision of several IEEE 802.15.4 channels of low data rate service. With some exceptions, unlicensed operation in this band is unique to North America, and regulations elsewhere preclude operation of unlicensed devices in this band.

To obtain economies of scale in product design, marketing, and distribution, and to enable applications that require roaming between different regulatory regions, it is desirable to employ a single band that is available on a nearly worldwide basis. The ideal band would be unlicensed, have sufficient width to enable the use of many channels, and be high enough in the spectrum so that relatively efficient antennas are possible without being so high that single-chip implementations employing low-cost integrated circuit processes are precluded.

The band selected as best meeting these requirements is the 2.4 GHz Industrial, Scientific, and Medical (ISM) band, which extends from 2400 MHz to 2483.5 MHz. This band, with very few exceptions, is available worldwide without licensing [7], [25], [26] and, with a wavelength of 12.25 cm, enables reasonably efficient yet physically small antennas. Further, it is compatible with modern silicon integrated circuit processes. Also, the wider bandwidth available enables the provision of multiple RF channels of relatively high data rate so that independent networks of similar devices can coexist without interfering with each other.

IEEE Std 802.15.4 was designed for regulatory compliance in each of the bands mentioned previously. For example, in the 868 MHz band, the product duty cycle is limited. To meet this requirement, the non-beacon mode of IEEE Std 802.15.4 may be used to minimize the duty cycle of devices in this band. In many cases, some type of spread spectrum operation is required for operation in unlicensed bands. The use of DSSS in IEEE Std 802.15.4 allows this regulatory requirement to be met, while also enabling a low-cost product implementation with good transmission range.

FREQUENCY BANDS AND DATA RATES

In response to the regulatory availability of regional or global bands for unlicensed operation, IEEE Std 802.15.4-2006 specifies technical objectives for operation the following bands:

- *868–868.6 MHz Band*: This unlicensed band is available in most European countries for a 20 kb/s BPSK and 100 kb/s O-QPSK service as well as an 250 kb/s Parallel Sequence Spread Spectrum (PSSS) service. IEEE Std 802.15.4 also refers to this band as the 868 MHz band.

- *902–928 MHz Band*: Some portions of this unlicensed band are available in North America, Australia, New Zealand, and some countries in South America for 40 kb/s BPSK and optional 250 kb/s O-QPSK service as well as an optional 250 kb/s PSSS service. IEEE Std 802.15.4 also refers to this band as the 915 MHz band (which represents the middle frequency of the band).

IEEE Std 802.15.4 868/915 MHz PHY requires that a compliant device be capable of operating in both the 868 MHz and 915 MHz bands. For the purposes of the standard, they are considered to be a single, contiguous band.

- *2.4000–2.4835 GHz Band:* This unlicensed band is available in most countries worldwide for the 250 kb/s O-QPSK service. This band is referred to as the 2.4 GHz band.

Amendments to IEEE Std 802.15.4-2006 revision have added new PHYs operating in other bands focused in new application areas (location awareness) or specialized in specific regional markets (China and Japan). These new PHY are:

> The 2006 revision of the original IEEE Std 802.15.4-2003 introduced two additional modulating options in the lower frequency band: O-QPSK and PSSS. The new PHY layer options provide a higher data rate compared to what was previously provided.

- *Amendment A*: Three Ultra-wide bands: 250–750 MHz, 3.1–4.8 GHz, and 6.0– 10.6 GHz designed to support location awareness applications in multiple operational regions, as well as one Chirp Spread Spectrum PHY collocated in the 2.4 GHz band.

- *Amendment C*: 779–787 MHz Band: This unlicensed band is available in China offering 250 kb/s O-QPSK and 250 kb/s MPSK services.

- *Amendment D*: 950–956 MHz Band: This unlicensed band is available for the Japanese market, offering a 100 kb/s GFSK and 20 kb/s BPSK service.

Because of its nearly worldwide availability, the 2.4 GHz PHY became the first choice for many IEEE Std 802.15.4 applications, especially those involving products that entail travel between regulatory regions. Even for non-mobile applications, the 2.4 GHz band offers advantages of scale, distribution, and marketing: a single product can be sold in multiple locations around the globe, without concern for the regulatory region in which it will be used. This drives down production costs, while eliminating the supply chain expense of tracking multiple products to multiple destinations.

However, for these same reasons, the 2.4 GHz band is used by many other services, from microwave ovens to WPANs and WLANs. The resulting congestion can be unacceptable for some applications and markets. To address

> Due to the slightly different application focus and technology complexity of IEEE Std 802.15.4a (Amendment A), this book does not cover the technology insights of the Ultra wide band PHY, including CSS. The actual Amendment A contains a large volume of introductory content for the interested reader.

this issue, IEEE Std 802.15.4 offers the possibility to use others bands such as the 868/915 MHz bands for Europe, the Chinese 780 MHz band, or the Japanese 950 MHz band. With this feature, system designers can use these regional bands to avoid the potentially crowded 2.4 GHz band. The regional bands could be a good design choice for applications such as utility meter reading that are inherently regional in nature and have limited device mobility. Similarly, diverse sensors in different application segments can use this feature.

As will be shown in the next chapters of this book, IEEE Std 802.15.4 was designed to effectively provide very high reliability in the event of interference cause by intentional (other radio systems) or unintentional radiators (microwave oven).

Data Rates

Due to the physical characteristics of each band and the regulations where they are used, IEEE Std 802.15.4 specifies different data rates and modulations for the bands used in the multiple PHYs defined in the standard. Table 4–1 shows the data rates (bit and symbol) and the modulation parameters specified in each band.

The mandatory base data rate in the 868 MHz band is 20 kb/s but implementers may choose adding either an optional higher data rate of 100 kb/s using O-QPSK modulation or 250 kb/s using PSSS modulation. Similarly, the mandatory base data rate in the 915 MHz is 40 kb/s with implementers able to provide an optional rate of 250 kb/s using either O-QPSK or PSSS modulation.

Any device operating in the 868/915 MHz bands must at least support a data rate of 20 kb/s or 40 kb/s using the specified modulation, respectively. Support for higher data rates using either modulation is optional.

When choosing the 950 MHz band the standard establishes that at least one of the two specified modulations needs to be implemented. To ensure interoperability in this band, it is recommended to implement both modulation schemes.

Table 4–1: IEEE Std 802.15.4 frequency band and modulation parameters

IEEE Std 802.15.4-2006

Band	Frequency Band	Bit Rate	Symbol Rate	DSSS Spreading Parameters Modulation	Chip Rate
868 MHz	868 - 868.6 MHz	20 kb/s	20 ksymbols/s	Binary Phase Shift Keying (BPSK)	300 kchip/s
		100 kb/s	25 ksymbols/s	Offset Quadrature Phase Shift Keying (O-QPSK)	400 kchip/s
		250 kb/s	12.5 ksymbols/s	Parallel Sequence Spread Spectrum (PSSS)	400 kchip/s
915 MHz	902 - 928 MHz	40 kb/s	40 ksymbols/s	Binary Phase Shift Keying (BPSK)	600 kchip/s
		250 kb/s	62.5 ksymbols/s	Offset Quadrature Phase Shift Keying (O-QPSK)	1 Mchip/s
		250 kb/s	50 ksymbols/s	Parallel Sequence Spread Spectrum (PSSS)	1.6 Mchip/s
2.4 GHz	2.4 - 2.4835 GHz	250 kb/s	62.5 ksymbols/s	Offset Quadrature Phase Shift Keying (O-QPSK)	2 Mchip/s

Amendment A

Band	Frequency Band	Bit Rate	Symbol Rate	Modulation
Ultra-Wide Band 1	250 – 750 MHz	110 kb/s	0.12 MHz	
		850 kb/s	0.98 MHz	
		1700 kb/s	1.95 MHz	
		6810 kb/s	7.8 MHz	
		27240 kb/s	15.6 / 31.2 MHz	
Ultra-Wide Band 2	3.1 GHz to 4.8 GHz	110 kb/s	0.12 MHz	Burst Position Modulation (BPM) with Binary Phase Shift Keying (BPSK)
		850 kb/s	0.98 MHz	
		1700 kb/s	1.95 MHz	
		6810 kb/s	7.8 MHz	
		27240 kb/s	15.6 / 31.2 MHz	
Ultra-Wide Band 3	6.0 GHz to 10.6 GHz	110 kb/s	0.12 MHz	
		850 kb/s	0.98 MHz	
		1700 kb/s	1.95 MHz	
		6810 kb/s	7.8 MHz	
		27240 kb/s	15.6 / 31.2 MHz	
2.4 GHz	2.4 - 2.4835 GHz	250 kb/s	166.67 ksymbols/s	Chirp Spread Spectrum (CSS)
		1000 kb/s		

Amendment C

Band	Frequency Band	Bit Rate	Symbol Rate	DSSS Spreading Parameters Modulation	Chip Rate
780 MHz	779 - 787 MHz	250 kb/s	62.5 ksymbols/s	Offset Quadrature Phase Shift Keying (O-QPSK) M-ary Phase Shift Keying (MPSK)	1000 kchip/s

Amendment D

Band	Frequency Band	Bit Rate	Symbol Rate	DSSS Spreading Parameters Modulation	Chip Rate
950 MHz	950 - 956 MHz	20 kb/s	20 ksymbols/s	Binary Phase Shift Keying (BPSK)	300 kchips/s
		100 kb/s	100 ksymbols/s	Gaussian Frequency Shift Keying (GFSK)	N/A

A more detailed explanation of the modulation parameters will be introduced in the following band specification sections.

CHANNEL ASSIGNMENT

IEEE Std 802.15.4-2006 and its Amendments A, C, and D define 16 channels in the 2450 MHz band, 30 channels in the 915 MHz band, 3 channels in the 868 MHz band, 14 overlapping chirp spread spectrum (CSS) channels in the 2450 MHz band, 16 channels in three UWB bands (500 MHz and 3.1 GHz to 10.6 GHz), 8 channels in the 780 MHz band, and 22 channels in 950 MHz band.

IEEE Std 802.15.4 uses a combination of channel numbers and channel pages to specify the operating frequency of a radio. There are a total of 32 possible pages; only seven are defined in the current standard, with the rest reserved for future expansion. Each page is logically divided into 27 channels, numbered 0 to 26. The pages are numbered 0 to 31. Page 0 contains 1 channel in the 868 MHz band, 10 channels in the 915 MHz band, and 16 channels in the 2.4 GHz band. The center frequencies of the channels for page 0 are shown in Table 4–2.

> The concept of channel pages was introduced in the 2006 revision of the standard to accommodate the new physical layer options and to allow for future expansion. The original 27 channels specified in IEEE Std 802.15.4-2003 are now accessed through page 0, while each of the new modulation mechanisms operate on their respective individual channel pages.

Channel page 1 contains the channel assignment for an optional PSSS high data-rate service in the 868/915 MHz frequency bands while channel page 2 specifies the channel assignment for an optional O-QPSK physical layer service in the same bands. Both pages 1 and 2 shown in Table 4–3 contain one channel for operation in the 868 MHz band and 10 channels for operation in the 915 MHz band, with the remaining 15 channels being reserved.

Table 4–2: IEEE Std 802.15.4 channel assignment for page 0

	Channel	Center Frequency (MHz)	Availability
868 MHz Band	0	868.3	
915 MHz Band	1	906	
	2	908	
	3	910	
	4	912	
	5	914	
	6	916	
	7	918	
	8	920	
	9	922	
	10	924	
2.4 GHz Band	11	2405	
	12	2410	
	13	2415	
	14	2420	
	15	2425	
	16	2430	
	17	2435	
	18	2440	
	19	2445	
	20	2450	
	21	2455	
	22	2460	
	23	2465	
	24	2470	
	25	2475	
	26	2480	

Table 4–3: IEEE Std 802.15.4 channel assignment for pages 1 and 2

	Channel	Center Frequency (MHz)	Availability
868 MHz Band	0	868.3	
915 MHz Band	1	906	
	2	908	
	3	910	
	4	912	
	5	914	
	6	916	
	7	918	
	8	920	
	9	922	
	10	924	

Channel pages 3 and 4 were created to support Amendment A of IEEE Std 802.15.4 (named 802.15.4a). Page 3 contains the channel assignment for the optional CSS PHY made up of 14 channels as shown in Table 4–4. Similarly, page 4 contains the channel assignment for the optional ultra-wide band PHYs composed of 16 channels grouped into three UWB bands named sub-GHz band, Low Band, and High Band. Detail of the channel assignment for page 4 is shown in Table 4–5.

Table 4–4:   IEEE Std 802.15.4a channel assignment, page 3

	Channel	Center Frequency (MHz)
	0	2412
	1	2417
	2	2422
	3	2427
	4	2432
2.4 GHz	5	2437
Band	6	2442
	7	2447
CSS	8	2452
	9	2457
	10	2462
	11	2467
	12	2472
	13	2484
	14-27	Unassigned

Channel page 5 was created to support Amendment C of IEEE Std 802.15.4 or IEEE Std 802.15.4c. This page contains the channel assignment for its four O-QPSK channels, as well as its four MPSK channels. Similarly, Channel page 6 was created to support Amendment D – IEEE Std 802.15.4d. This page contains the channel assignment for 10 BPSK channels and 12 GFSK channels. Details on pages 5 and 6 are shown in Table 4–6 and Table 4–7, respectively.

Table 4–5: IEEE Std 802.15.4a channel assignment, page 4

	Channel	Center Frequency (MHz)	Bandwidth (MHz)
UWB 1 SubGHz 250 – 750 MHz	0	499.2	499.2
UWB 2 Low Band 3.1 GHz to 4.8 GHz	1	3494.4	499.2
	2	3993.6	499.2
	3	4492.8 *	499.2
	4	3993.6	1331.2
UWB 3 High Band 6.0 GHz to 10.6 GHz	5	6489.6	499.2
	6	6988.8	499.2
	7	6489.6	1081.6
	8	7488	499.2
	9	7987.2*	499.2
	10	8486.4	499.2
	11	7987.2	1331.2
	12	8985.6	499.2
	13	9484.8	499.2
	14	9984	499.2
	15	9484.8	1354.97
	16-27	Unassigned	N/A

Table 4–6: IEEE Std 802.15.4c channel assignment, page 5

	Channel	Center Frequency (MHz)	Availability
780 MHz Band O-QPSK	0	780	
	1	782	
	2	784	
	3	786	
780 MHz Band MPSK	4	780	
	5	782	
	6	784	
	7	786	
	8-27	Unassigned	

Table 4–7: **IEEE Std 802.15.4d channel assignment, page 6**

	Channel	Center Frequency (MHz)	Availability
950 MHz Band BPSK	0	951.2	
	1	951.8	
	2	952.4	
	3	953	
	4	953.6	
	5	954.2	
	6	954.8	
	7	955.4	
	8	954.4	
	9	954.6	
950 MHz Band GFSK	10	951.1	
	11	951.5	
	12	951.9	
	13	952.3	
	14	952.7	
	15	953.1	
	16	953.5	
	17	953.9	
	18	954.3	
	19	954.7	
	20	955.1	
	21	955.5	
	22-27	Unassigned	

NEW OPTIONAL PHYSICAL LAYERS

Although the lower data rates in the 868/915 MHz band have the advantage of improved range performance, the result is lower data throughput with respect to the 2.4 GHz–band data rate. This can be a particular challenge for larger networks operating in these bands. For the European 868 MHz band this challenge is emphasized since the data throughput is limited not only by the 20 kb/s data rate (itself a result of the limited available bandwidth), but also by the regulatory duty cycle limitation. A device operating in the 868 MHz band is allowed to transmit with only a 1% duty cycle. To alleviate this disadvantage, the 2006 revision of the original standard added two new physical layer options providing higher data rates in the lower frequency bands. The two optional PHYs provide more choices to the implementer and user. They also allow original equipment manufacturers certain performance trade-offs based on application requirements when choosing a solution. The two new optional physical layers include an O-QPSK PHY with a data rate of 100 kb/s at 868 MHz and 250 kb/s at 915 MHz, and also a PSSS PHY with a data rate of 250 kb/s in both of these bands. However, for backward compatibility with existing devices and for interoperability among new devices, the standard requires implementers of either option to also support the original 868/915 MHz physical layer that was introduced by the original 2003 version of the standard.

The optional O-QPSK PHY specification for the 868/915 MHz band is a derivative of the modulation used also in the 2.4 GHz band, allowing implementers to share design similarities with existing radios and enable a potential three-band transceiver. The optional PSSS PHY offers the same data rate at both low-band frequencies and uses a more complex modulation scheme but has the advantage of providing improved multipath performance.

In addition, as introduced in the previous section, Amendment A added four ultra-wideband physical layers, Amendment C added two optional physical layers designed to serve the Chinese market, and finally Amendment D added two additional optional physical layers designed to serve the Japanese market.

> The 2006 revision of IEEE Std 802.15.4 introduced two new physical layer options for the lower frequency bands that increase the throughput to equivalent levels of the 2.4 GHz band.

PHY BIT LEVEL COMMUNICATION

The physical layer of the IEEE Std 802.15.4 protocol is responsible for establishing the RF link between two devices. It also provides for bit modulation, demodulation, and synchronization between the transmitter and receiver. Finally, the PHY also handles the packet level synchronization.

IEEE Std 802.15.4 specifies several data rates that can be used in the different bands defined. A device compliant with the 868/915 MHz PHY layer specifications of IEEE Std 802.15.4-2006 must also support the two lowest rates of 20kb/s and 40kb/s using the BPSK modulation scheme. Optionally, the device may also provide a 100 kb/s or 250 kb/s data service using either O-QPSK modulation or a 250 kb/s data service using PSSS modulation. The 2.4 GHz PHY layer provides a data rate of 250 kb/s using an O-QPSK modulation scheme with an m-ary quasi-orthogonal modulation technique. In addition, Amendment A defined a 2.4 GHz Chirp Spread Spectrum PHY offering data rates of 250 kb/s and 1000 kb/s. This same amendment defined three Ultra-Wide band PHYs offering five distinct data rates of 110 kb/s, 850 kb/s, 1,700 kb/s, 6,810 kb/s, and 27,240 kb/s.

Amendments C and D defined two additional optional PHYs in the 780 MHz and 950 MHz bands respectively. A device compliant with the 780 MHz PHY will support a data rate of 250 kb/s using O-QPSK or MPSK. Similarly, a device compliant with the 950-MHz PHY will support data rates of 20 kb/s and 100 kb/s using BPSK or GFSK.

2.4 GHz Band Specification

The 16-ary quasi-orthogonal modulation technique specified in the IEEE Std 802.15.4 2.4 GHz PHY layer is a method of data modulation that uses a particular 32-chip, pseudo-random sequence to represent four bits and simultaneously accomplish the spreading modulation. The data modulation is performed by means of cyclic rotation and/or conjugation (inversion of chips with odd indices) of the sequence. The pseudo-random sequence is started in different places, depending on the modulating data, transmitting four bits in each symbol period.

Although five bits could be transmitted by the choice of 32 chips, four were chosen for the 2.4 GHz PHY to minimize implementation complexity. The transmitted 32-chip pseudo-random sequence is allowed to start only at every fourth chip of the sequence. Symbols 0–7 represent cyclic shifts in multiples of four chips. Symbols 8–15 use the same shifts as symbols 0–7, respectively, but use the conjugated sequence (i.e., the odd-indexed chips are inverted).

The IEEE Std 802.15.4 2.4 GHz PHY layer specifies a symbol rate of 62.5 ksymbols per second with four bits in each symbol; therefore, 250 kb/s service is attained. The 32-chip pseudo-random sequence to be transmitted is split between the orthogonal I and Q channels of the O-QPSK modulator, with the even-indexed chips placed on the I channel and the odd-indexed chips placed on the Q channel. A one-half chip delay is placed in the Q channel, creating the offset for O-QPSK. Because 32 (now complex) chips are transmitted in one symbol time (16 μs), the overall chip rate is 2 Mc/s. The chip rate in either I or Q channel, however, is 1 Mc/s.

Initially two distinct pseudo-noise (PN) sequences were considered for the I and Q channels of the quasi-orthogonal modulation in the O-QPSK, but the two were later combined to use a single PN sequence separated into even and odd bits between the I and Q channels. The advantage of a single PN sequence lies in that a single correlator is required that can be shared between the I and Q channels, thereby reducing implementation complexity.

The bit level processing consists of assembling four bits into a symbol, converting that symbol to a cyclically rotated 32-chip sequence as shown in Table 4–8, and modulating that chip sequence on the I or Q channel, respectively. The process is diagrammed in Figure 4–1.

Because 32 chips are transmitted in one symbol time (16 μs), the overall chip rate is 2 Mc/s, and the length of a chip is Tc = 0.5 μs. Successive chips in each of the I and Q channels start every 2 Tc.

Table 4–8: Symbol to chip mapping

Data Symbol (decimal)	Data Symbol (binary) (b3 b2 b1 b0)	Chip Values (c0 c1... C30 C31)
0	0000	11011001110000110101001000101110
1	0001	11101101100111000011010100100010
2	0010	00101110110110011100001101010010
3	0011	00100010111011011001110000110101
4	0100	01010010001011101101100111000011
5	0101	00110101001000101110110110011100
6	0110	11000011010100100010111011011001
7	0111	10011100001101010010001011101101
8	1000	10001100100101100000011101111011
9	1001	10111000110010010110000001110111
10	1010	01111011000110010010110000000111
11	1011	01110111101110001100100101100000
12	1100	00000111011110111000110010010110
13	1101	01100000011101111011100011001001
14	1110	10010110000001110111101110001100
15	1111	11001001011000000111011110111000

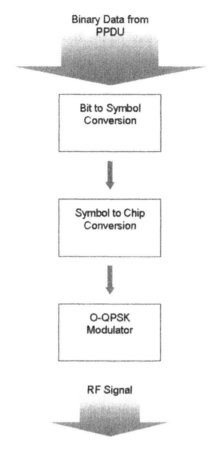

Figure 4–1: 2.4 GHz modulation and spreading

The resulting combination of complex modulation, including the offset between I and Q channels is shown in Figure 4–2.

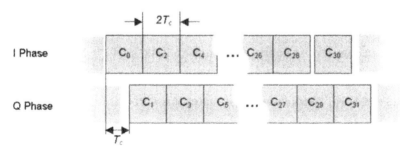

Figure 4–2:   O-QPSK chip offsets

868/915 MHz Band Specification

The 868/915 MHz PHY layer based on the 2006 revision allows for multiple data rates using different modulation schemes thus providing OEMs with the opportunity to trade off certain features and performance characteristics, depending on application requirements. To allow for backward compatibility, all devices compliant with the 868/915 MHz PHY layer specifications of IEEE Std 802.15.4-2006 are required to support the 20 kb/s and 40 kb/s data services of the 2003 version of the standard using the respective spreading parameters. This base mode of operation allows new devices to be able to communicate with IEEE Std 802.15.4-2003 compliant devices that are already deployed in the field and allows new devices that employ the new PHY layer options to interoperate. New applications or deployments without legacy devices can take advantage of the higher data rates and performance characteristics that the new physical layer options provide.

Required BPSK Mode Specification

The base mode of the 868/915 MHz PHY layer uses DSSS employing BPSK, providing a data rate of 20 kb/s in the 868 MHz band and 40 kb/s in the 915 MHz band. A complex signal path is not required for this mode.

The 868/915 MHz PHY specifies differential encoding of the transmitted data bits. If the raw data bit is 0, the BPSK data bit is transmitted in the same phase as the previous BPSK data bit. If the raw data bit is 1, the BPSK data bit is transmitted in phase opposite to the previous BPSK bit.

The 868/915 MHz PHY specifies conventional DSSS, in which a single, 15-chip, pseudo-random sequence is transmitted in a symbol period to represent a 1, and the inverse of the sequence is transmitted to represent a 0. This process is illustrated in Figure 4–3. The chip rate is specified as 300 kc/s in the 868 MHz band, for a data rate of 20 kb/s; in the 915 MHz band, the specified chip rate of 600 kc/s enables a data rate of 40 kb/s. The reception process used to recover the transmitted data is shown in Figure 4–4.

The 868/915 MHz modulation and spreading process is illustrated in Figure 4–5.

The BPSK modulation uses a different PN sequence, but it shares that sequence between the two lower data rates. The difference in data rates is due to the difference in chip rates, but the number of chip bits per data bit and the bit pattern are the same between the two.

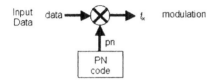

		Band	
		868 MHz	915 MHz
T_b	bit period	50 us	25 us
R_b	bit rate	20 kb/s	40 kb/s
T_c	chip period	3.33 us	1.66 us
R_c	chip rate	300 kc/s	600 kc/s
N_c	chips per symbol	15	

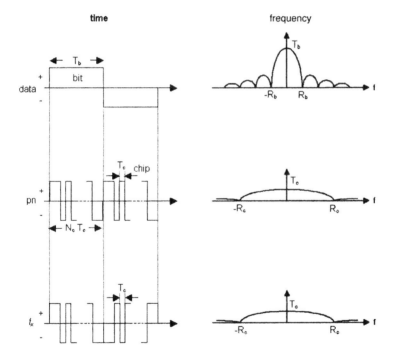

Figure 4–3: DSSS modulation

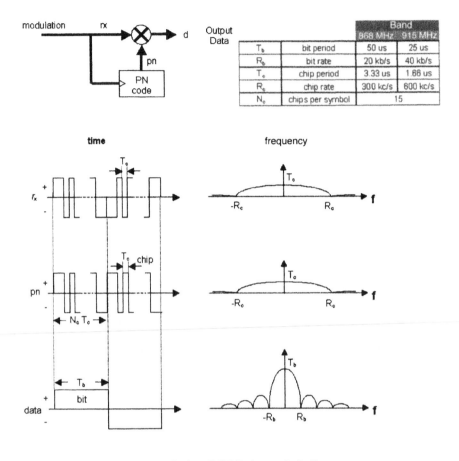

		Band	
		868 MHz	915 MHz
T_b	bit period	50 us	25 us
R_b	bit rate	20 kb/s	40 kb/s
T_c	chip period	3.33 us	1.66 us
R_c	chip rate	300 kc/s	600 kc/s
N_c	chips per symbol	15	

Figure 4–4: DSSS demodulation

Optional O-QPSK Mode Specification

The optional O-QPSK modulation scheme for the lower frequency bands is a derivative of the modulation used in the 2.4 GHz physical layer, and provides 100 kb/s data service at 868 MHz, and 250 kb/s data service at the 915 MHz band. It allows a significant improvement in data throughput over the base 868/915 MHz band BPSK mode while allowing implementers to leverage design similarities with the high band PHY. It employs a 16-ary quasi-orthog-onal modulation scheme with a nearly orthogonal, 16-chip, pseudo-random noise sequence to modulate four data bits into each symbol. Identical to the 2.4 GHz PHY, the data modulation is performed by a cyclic rotation and/or

The new optional O-QPSK PHY offers increased data throughput compared to the base mode while allowing implementers to share design similarities with existing transceivers operating in the 2.4 GHz band.

conjugation of the chip sequence. Symbols 0–7 represent cyclic shifts in multiples of two chips. Symbols 8–15 use the same shifts as symbols 0–7, respectively, but use the conjugated sequence (i.e., the odd-indexed chips are inverted). The chip sequences used for this modulation scheme are shown in Table 4–9.

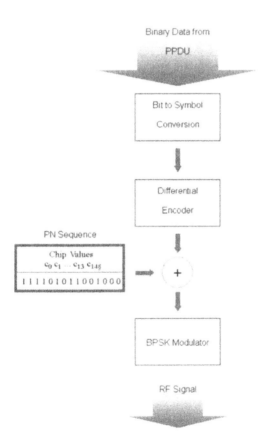

Figure 4–5: 868/915 MHz modulation and spreading

Table 4–9: Symbol to chip mapping for optional O-QPSK modulation

Data Symbol (decimal)	Data Symbol (binary) (b3 b2 b1 b0)	Chip Values (c0 c1 ... c14 c15)
0	0000	0 0 1 1 1 1 1 0 0 0 1 0 0 1 0 1
1	0001	0 1 0 0 1 1 1 1 1 0 0 0 1 0 0 1
2	0010	0 1 0 1 0 0 1 1 1 1 1 0 0 0 1 0
3	0011	1 0 0 1 0 1 0 0 1 1 1 1 1 0 0 0
4	0100	0 0 1 0 0 1 0 1 0 0 1 1 1 1 1 0
5	0101	1 0 0 0 1 0 0 1 0 1 0 0 1 1 1 1
6	0110	1 1 1 0 0 0 1 0 0 1 0 1 0 0 1 1
7	0111	1 1 1 1 1 0 0 0 1 0 0 1 0 1 0 0
8	1000	0 1 1 0 1 0 1 1 0 1 1 1 0 0 0 0
9	1001	0 0 0 1 1 0 1 0 1 1 0 1 1 1 0 0
10	1010	0 0 0 0 0 1 1 0 1 0 1 1 0 1 1 1
11	1011	1 1 0 0 0 0 0 1 1 0 1 0 1 1 0 1
12	1100	0 1 1 1 0 0 0 0 0 1 1 0 1 0 1 1
13	1101	1 1 0 1 1 1 0 0 0 0 0 1 1 0 1 0
14	1110	1 0 1 1 0 1 1 1 0 0 0 0 0 1 1 0
15	1111	1 0 1 0 1 1 0 1 1 1 0 0 0 0 0 1

The symbol rate for the optional O-QPSK PHY is 25 ksymbol/s with a chip rate of 400 kchip/s at the 868 MHz band and 62.5 ksymbol/s with a chip rate of 1 Mchip/s at the 915 MHz band resulting in the previously mentioned 100 kb/s and 250 kb/s data services respectively. Before transmission, the symbol is split between the I and Q channels of the modulator. The even-indexed chips are put into the I-channel while the odd-indexed chips are put in the Q-channel. The transmission of the I-channel and Q-channel are offset by half a chip period.

The modulation process is identical to the one of the 2.4 GHz physical layer shown in Figure 4–1 on page 57. The resulting complex modulation and its offset is shown in Figure 4–6.

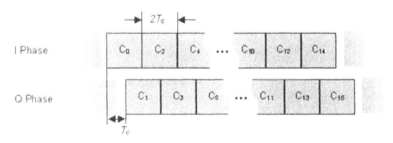

Figure 4–6: O-QPSK chip offsets for optional 868/915 MHz PHY

Optional PSSS Mode Specification

The optional Parallel Sequence Spread Spectrum (PSSS) physical layer uses a multicode modulation scheme [22] based on amplitude shift keying (ASK). The advantage of this PSSS mode PHY is that it provides an improved multi-path performance compared to the

> The new optional PSSS PHY offers increased data throughput compared to the base mode and improved performance character-istics as result of the advanced multicoded modulation scheme.

other two low-band physical layers while also achieving the same data rate available at the 2.4 GHz PHY in the 868 MHz and 915 MHz bands. However, the drawback is a slightly more complex transceiver design compared to the other options.

The PSSS mode PHY uses a 31-chip base sequence. For the 868 MHz band, this sequence is rotated by 1.5 chips. Then a one chip cyclic extension is added, which results in 20 nearly orthogonal pseudo-random sequences that are 64 half-chips long. The PSSS code table used for the 868 MHz band is show in Table 4–10. For the 915 MHz band, the sequence is cyclically shifted by six chips with a one chip extension added. This operation results in five nearly orthogonal pseudo-random sequences that are 32 chips long. The PSSS code sequence used in the 915 MHz band is shown in Table 4–11. The modulation process consists taking the binary data of the PHY header and PHY payload and converting 20 of these data bits for the 868 MHz band and five of these data bits for the 915 MHz band into symbols. The last symbol is padded with "0" bits if less than the required number of bits remain in order to fill the last symbol. The individual bits of the symbol are converted to bipolar levels where a 1 bit becomes a +1 and a 0 bit becomes a –1. This is followed by the symbol-to-chip conversion and then applying the ASK modulation as shown in Figure 4–7. For this process the bipolar version of the first bit of the symbol is multiplied by the first sequence in the respective PSSS code table, the bipolar version of the second bit of the symbol is multiplied with the second sequence in the PSSS code table, and so on. The effect is that, depending on the data bit, the respective PSSS code sequence is either inverted or not.

Table 4–10: PSSS code table for 868 MHz band

Sequence Number	Chip Number																															
	0	1	2	3	4	5	6	7	8	9	10	11	12	13	14	15	16	17	18	19	20	21	22	23	24	25	26	27	28	29	30	31
0	-1	-1	-1	-1	1	-1	-1	-1	1	-1	1	1	-1	-1	1	1	1	1	-1	-1	-1	1	1	1	1	1	1	-1	1	1	1	-1
1	1	1	-1	1	-1	1	-1	-1	-1	-1	-1	1	-1	-1	1	1	1	1	-1	-1	1	1	1	1	-1	-1	-1	1	1	-1	1	1
2	-1	-1	1	1	1	1	1	-1	1	-1	1	-1	1	-1	-1	1	1	1	-1	1	-1	-1	1	1	-1	-1	1	1	1	-1	-1	-1
3	1	1	1	1	1	-1	-1	-1	1	-1	1	1	1	-1	1	1	1	-1	-1	1	-1	-1	1	-1	1	-1	1	1	1	1	-1	1
4	1	-1	1	1	-1	-1	1	1	1	1	1	-1	-1	-1	1	1	-1	1	1	1	-1	1	-1	1	-1	-1	-1	-1	1	-1	-1	1

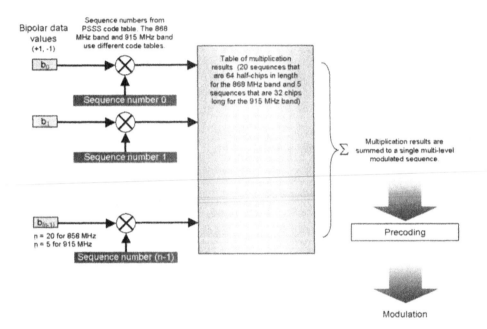

Figure 4–7: Symbol-to-chip mapping for PSSS mode PHY

The products of these multiplications (20 multiplication results when operating in the 868 MHz band and five multiplication results when operating in the 915 MHz band) are summed to a single multilevel modulation sequence. A two-step precoding process follows to eliminate the DC content of the multilevel modulated sequence and to normalize the sequence. To transmit the sequence over the medium, it is modulated onto a carrier using amplitude shift keying (ASK).

Table 4–11: PSSS code table for 915 MHz band

Sequence Number	Chip Number																															
	0	1	2	3	4	5	6	7	8	9	10	11	12	13	14	15	16	17	18	19	20	21	22	23	24	25	26	27	28	29	30	31
0																																
1																																
2																																
3																																
4																																
5																																
6																																
7																																
8																																
9																																
10																																
11																																
12																																
13																																
14																																
15																																
16																																
17																																
18																																
19																																

Half-chip Number: 0 1 2 3 4 5 6 7 8 9 10 11 12 13 14 15 16 17 18 19 20 21 22 23 24 25 26 27 28 29 30 31 32 33 34 35 36 37 38 39 40 41 42 43 44 45 46 47 48 49 50 51 52 53 54 55 56 57 58 59 60 61 62 63

The synchronization header is transmitted using BPSK modulation at the same chip rate and pulse shaping used for the PHY header and payload. The modulation and spreading functions for the PSSS mode PHY are shown in Figure 4–8.

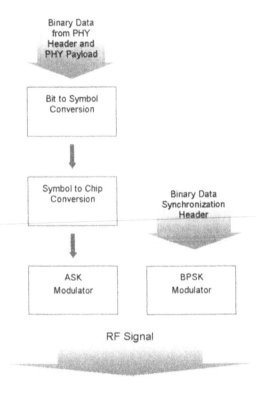

Figure 4–8: 868/915 MHz PSSS PHY modulation and spreading

780 MHz Band Specification

The 780 MHz PHY specifies a unique data rate of 250 kb/s using two possible modulation schemes: O-QPSK or MPSK.

The 780 MHz O-QPSK physical layer uses the same 16-ary quasi-orthogonal modulation and spreading used in the 2.4 GHz band specification, as well as symbol-to-chip mapping and bit timing. With this physical layer, radio chip designers can use existing 2.4 GHz designs in the 780 Mhz band.

Similar to the O-QPSK PHY, the 780 MHz MPSK physical layer employs a 16-ary orthogonal modulation technique where for each data symbol period, four information bits are used to map one of 16 orthogonal pseudo-random sequences to be transmitted. The PN sequences for successive data symbols are then concatenated, and the aggregate chip phase is modulated onto the carrier using PSK. Table 4–12 shows the symbol to chip mapping for the MSPK modulation.

Table 4–12: Symbol-to-chip mapping for MPSK

Data Symbol (decimal)	Data Symbol (binary) (b3 b2 b1 b0)	Chip Values (c0 c1 ... c30 c31)
0	0000	0 $\frac{\pi}{4}$ $\frac{\pi}{16}$ $\frac{9\pi}{16}$ π $-\frac{7\pi}{16}$ $\frac{\pi}{4}$ $-\frac{15\pi}{16}$ 0 $-\frac{15\pi}{16}$ $\frac{\pi}{4}$ $\frac{7\pi}{16}$ π $\frac{9\pi}{16}$ $\frac{\pi}{4}$ $\frac{\pi}{16}$
1	0001	$\frac{\pi}{16}$ 0 $\frac{\pi}{4}$ $\frac{\pi}{16}$ $\frac{9\pi}{16}$ π $-\frac{7\pi}{16}$ $\frac{\pi}{4}$ $-\frac{15\pi}{16}$ 0 $-\frac{15\pi}{16}$ $\frac{\pi}{4}$ $\frac{7\pi}{16}$ π $\frac{9\pi}{16}$ $\frac{\pi}{4}$
2	0010	$\frac{\pi}{4}$ $\frac{\pi}{16}$ 0 $\frac{\pi}{4}$ $\frac{\pi}{16}$ $\frac{9\pi}{16}$ π $-\frac{7\pi}{16}$ $\frac{\pi}{4}$ $-\frac{15\pi}{16}$ 0 $-\frac{15\pi}{16}$ $\frac{\pi}{4}$ $\frac{7\pi}{16}$ π $\frac{9\pi}{16}$
3	0011	$\frac{9\pi}{16}$ $\frac{\pi}{4}$ $\frac{\pi}{16}$ 0 $\frac{\pi}{4}$ $\frac{\pi}{16}$ $\frac{9\pi}{16}$ π $-\frac{7\pi}{16}$ $\frac{\pi}{4}$ $-\frac{15\pi}{16}$ 0 $-\frac{15\pi}{16}$ $\frac{\pi}{4}$ $-\frac{7\pi}{16}$ π
4	0100	π $\frac{9\pi}{16}$ $\frac{\pi}{4}$ $\frac{\pi}{16}$ 0 $\frac{\pi}{4}$ $\frac{\pi}{16}$ $\frac{9\pi}{16}$ π $-\frac{7\pi}{16}$ $\frac{\pi}{4}$ $-\frac{15\pi}{16}$ 0 $-\frac{15\pi}{16}$ $\frac{\pi}{4}$ $-\frac{7\pi}{16}$
5	0101	$-\frac{7\pi}{16}$ π $\frac{9\pi}{16}$ $\frac{\pi}{4}$ $\frac{\pi}{16}$ 0 $\frac{\pi}{4}$ $\frac{\pi}{16}$ $\frac{9\pi}{16}$ π $-\frac{7\pi}{16}$ $\frac{\pi}{4}$ $-\frac{15\pi}{16}$ 0 $-\frac{15\pi}{16}$ $\frac{\pi}{4}$
6	0110	$\frac{\pi}{4}$ $-\frac{7\pi}{16}$ π $\frac{9\pi}{16}$ $\frac{\pi}{4}$ $\frac{\pi}{16}$ 0 $\frac{\pi}{4}$ $\frac{\pi}{16}$ $\frac{9\pi}{16}$ π $-\frac{7\pi}{16}$ $\frac{\pi}{4}$ $-\frac{15\pi}{16}$ 0 $-\frac{15\pi}{16}$
7	0111	$-\frac{15\pi}{16}$ $\frac{\pi}{4}$ $-\frac{7\pi}{16}$ π $\frac{9\pi}{16}$ $\frac{\pi}{4}$ $\frac{\pi}{16}$ 0 $\frac{\pi}{4}$ $\frac{\pi}{16}$ $\frac{9\pi}{16}$ π $-\frac{7\pi}{16}$ $\frac{\pi}{4}$ $-\frac{15\pi}{16}$ 0
8	1000	0 $-\frac{15\pi}{16}$ $\frac{\pi}{4}$ $-\frac{7\pi}{16}$ π $\frac{9\pi}{16}$ $\frac{\pi}{4}$ $\frac{\pi}{16}$ 0 $\frac{\pi}{4}$ $\frac{\pi}{16}$ $\frac{9\pi}{16}$ π $-\frac{7\pi}{16}$ $\frac{\pi}{4}$ $-\frac{15\pi}{16}$
9	1001	$-\frac{15\pi}{16}$ 0 $-\frac{15\pi}{16}$ $\frac{\pi}{4}$ $-\frac{7\pi}{16}$ π $\frac{9\pi}{16}$ $\frac{\pi}{4}$ $\frac{\pi}{16}$ 0 $\frac{\pi}{4}$ $\frac{\pi}{16}$ $\frac{9\pi}{16}$ π $-\frac{7\pi}{16}$ $\frac{\pi}{4}$
10	1010	$\frac{\pi}{4}$ $-\frac{15\pi}{16}$ 0 $-\frac{15\pi}{16}$ $\frac{\pi}{4}$ $-\frac{7\pi}{16}$ π $\frac{9\pi}{16}$ $\frac{\pi}{4}$ $\frac{\pi}{16}$ 0 $\frac{\pi}{4}$ $\frac{\pi}{16}$ $\frac{9\pi}{16}$ π $-\frac{7\pi}{16}$
11	1011	$-\frac{7\pi}{16}$ $\frac{\pi}{4}$ $-\frac{15\pi}{16}$ 0 $-\frac{15\pi}{16}$ $\frac{\pi}{4}$ $-\frac{7\pi}{16}$ π $\frac{9\pi}{16}$ $\frac{\pi}{4}$ $\frac{\pi}{16}$ 0 $\frac{\pi}{4}$ $\frac{\pi}{16}$ $\frac{9\pi}{16}$ π
12	1100	π $-\frac{7\pi}{16}$ $\frac{\pi}{4}$ $-\frac{15\pi}{16}$ 0 $-\frac{15\pi}{16}$ $\frac{\pi}{4}$ $-\frac{7\pi}{16}$ π $\frac{9\pi}{16}$ $\frac{\pi}{4}$ $\frac{\pi}{16}$ 0 $\frac{\pi}{4}$ $\frac{\pi}{16}$ $\frac{9\pi}{16}$
13	1101	$\frac{9\pi}{16}$ π $-\frac{7\pi}{16}$ $\frac{\pi}{4}$ $-\frac{15\pi}{16}$ 0 $-\frac{15\pi}{16}$ $\frac{\pi}{4}$ $-\frac{7\pi}{16}$ π $\frac{9\pi}{16}$ $\frac{\pi}{4}$ $\frac{\pi}{16}$ 0 $\frac{\pi}{4}$ $\frac{\pi}{16}$
14	1110	$\frac{\pi}{16}$ $\frac{9\pi}{16}$ π $-\frac{7\pi}{16}$ $\frac{\pi}{4}$ $-\frac{15\pi}{16}$ 0 $-\frac{15\pi}{16}$ $\frac{\pi}{4}$ $-\frac{7\pi}{16}$ π $\frac{9\pi}{16}$ $\frac{\pi}{4}$ $\frac{\pi}{16}$ 0 $\frac{\pi}{4}$
15	1111	$\frac{\pi}{4}$ $\frac{\pi}{16}$ $\frac{9\pi}{16}$ π $-\frac{7\pi}{16}$ $\frac{\pi}{4}$ $-\frac{15\pi}{16}$ 0 $-\frac{15\pi}{16}$ $\frac{\pi}{4}$ $-\frac{7\pi}{16}$ π $\frac{9\pi}{16}$ $\frac{\pi}{4}$ $\frac{\pi}{16}$ 0

950 MHz Band Specification

The 950 MHz PHY specifies two data rates, 20 kb/s using direct sequence spread spectrum with BPSK and a 100 kb/s using GFSK.

The 950 MHz BPSK physical layer was designed to use the 868 MHz BPSK PHY with the only difference of defining a more stringent power spectral density. The 950 MHz GFSK PHY modulate bit sequences onto a GFSK carrier with modulation index equal to 1 and using a Gaussian filter with BT of 0.5 at a rate of 100 kb/s.

RADIO CHARACTERISTICS

IEEE Std 802.15.4 radio specification was designed to allow the implementation of low-cost digital integrated circuit designs. Most of the technical radio specification can be considered relaxed with respect to other radio technologies. The following paragraphs present some of the characteristics of IEEE Std 802.15.4 radio.

Power Output

IEEE Std 802.15.4 provides for a wide transmitter output power range, but the device must be capable of transmitting –3 dBm. The upper limit of power output is designated by the regulatory agency associated with the local use. For example, in the United States, some services using DSSS in the 2.4 GHz band are allowed up to 1 W of transmitter power [25]; however, in Europe the limit is 100 mW in the same band [7].

Sensitivity

IEEE Std 802.15.4 specifies that the receiver must be capable of correctly decoding a signal with an input power of –85 dBm or less in the 2.4 GHz band as well as the optional O-QPSK, PSSS, and GFSK PHYs in the 780, 868/915, and 950 MHz bands respectively. In the lower frequency bands where BPSK is used, the receiver must be capable of correctly decoding a signal with an input power of –92 dBm or less. Better sensitivity is not prohibited.

Antennas that have a constant effective aperture exist; such antennas have gain that increases as the frequency of operation increases. An example of such an antenna is the parabolic dish antenna. Users of a communication link employing parabolic dish antennas of fixed dimensions on both transmitter and receiver would notice a range extension as the frequency of operation was increased— just the opposite effect of that the users employing dipole antennas see.

Range

In free-space, the path loss between a transmitter and a receiver is dependent solely on the distance between them, and is independent of the frequency used. There is no inherent difference in range between IEEE Std 802.15.4 frequency bands. The type of antenna used, however, can give the illusion of a frequency-dependent path loss: A constant-gain antenna, such as a half-wave dipole, has an effective aperture (area) that decreases as the frequency of operation increases. (The dipole gets physically smaller as the frequency increases, leading to a reduction of its effective area.) At higher frequencies, the dipole therefore intercepts less of the transmitted radiation and produces less received signal at its terminals. Users of a communication link employing dipole antennas on both transmitter and receiver would therefore notice a range reduction as the frequency of operation increased, although the effect was due to the antennas employed, rather than an increase in path loss.

The previous notwithstanding, many implementations of IEEE Std 802.15.4 are expected to employ constant-gain (dipole) antennas. For these implementations, in free space the ratio of the power available at the terminals of a receiving antenna to the power supplied to the terminals of a transmitting antenna is given by the Friis model:

$$\frac{P_R}{P_T} = G_T G_R \left(\frac{\lambda}{4\pi d}\right)^2 = G_T G_R \left(\frac{c}{4\pi f d}\right)^2 \qquad \text{EQ 4-1}$$

where P_R and P_T are the power values at the receiving antenna and transmitting antenna, respectively, (in watts); G_R and G_T are the power gains of the receiving antenna and transmitting antenna, respectively; λ is the wavelength (in meters); f is the frequency (in Hertz); d is the distance (in meters), and c is the speed of light (in meters/second).

Continuing under the assumption of constant-gain antennas, this same equation can be expressed as a basic transmit-antenna-terminal-to-receive-antenna-terminal loss L_B in decibel form, with the appropriate substitutions for the constants as:

$$L_b(dB) = 32.44 + 20\log_{10} df_{\text{MHZ}} + 20\log_{10} d_{km} \qquad \textbf{EQ 4-2}$$

in which the loss L_B includes contributions from both the antenna effective area (which decreases with frequency), and the path loss itself (which is constant with frequency). Here, since we assume constant-gain antennas, we have also normalized G_R and G_T to 1. For further simplification, we treat the 780, 868, 915, and 950 MHz bands as approximately 1 GHz and the 2.4 GHz band as approximately 2 GHz so that the losses can be expressed as:

$$L_{b \text{ @ 1GHz}} = 92 + 20\log_{10} d_{km} \qquad \textbf{EQ 4-3}$$

$$L_{b \text{ @ 2GHz}} = 98 + 20\log_{10} d_{km} \qquad \textbf{EQ 4-4}$$

For the IEEE Std 802.15.4 PHYs specifying a sensitivity of at least –92 dBm in the vicinity of 1 GHz, for a transmitter rated at 0 dBm, the maximum free space range is approximately 1 km ($\log_{10}[1]=0$). Under the same conditions for PHYs specifying a sensitivity of –85 dBm, the maximum free space range is approximately 450 meters.

For the higher 2.4 GHz band, the receiver must exhibit a sensitivity of –85 dBm; for a transmitter rated at 0 dBm, similar calculations will show a maximum free space range of approximately 220 meters. Note again that the range difference is due entirely to the type of antenna used.

Be aware that the previous derivations are using the free-space radio propagation model. These calculations are ideal; make sure you finish reading this section, where a real-life model is used.

These ranges are for free space, using perfectly matched, unity-gain antennas, without interference, and represent the maximum theoretical distances. The free-space model does not consider several environment parameters that impair the RF propagation, including wave reflection, diffraction, and scattering. A commonly used model that approximates the propagation behavior in RF channels in real environments is the log-normal shadowing model. Figure 4–9 illustrates a typical indoor propagation effect and the variance in range obtained due to the effect of shadowing (transmitting a message with an output power of 0 dBm). The graphs show the received isotropic power vs. distance at 915 MHz and 2.4 GHz, respectively. As it can be seen, the distance decreases with increasing frequency, indicating the assumption of constant-gain antennas in this analysis. Similarly, the variability shown is a function of the characteristic of the wireless channel. The path loss coefficient in indoor environments can vary between $n = 2$ and $n = 4$ typically. Furthermore, the indoor variance in the received power ranges from $\sigma = 3$ to $\sigma = 11$.

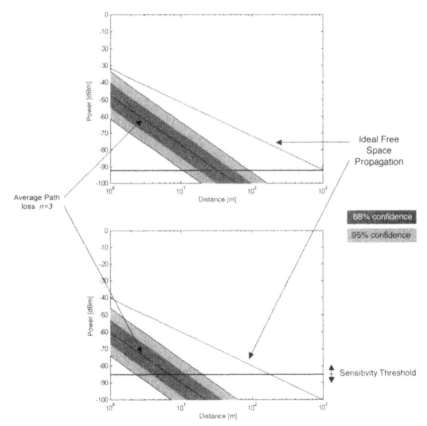

Figure 4–9: Typical indoor range (0 dBm transmission) with normal distribution (= 7dB): (a) 915 MHz, (b) 2.4 GHz

Receiver Selectivity

The DSSS signal is, by definition, a wideband variation of the modulated data signal. The DSSS spreading provides the communications benefits already discussed, and it provides a relaxed RF selectivity requirement. In addition, IEEE Std 802.15.4 channels are widely spaced (5 MHz in the 2.4 GHz band) relative to their own bandwidth (3 MHz null-to-null; about 1.5 MHz noise bandwidth in the 2.4 GHz band). With the 0 dB adjacent channel requirement, as well as the relaxed requirement for channels further away, little selectivity is needed by the implementer to achieve the requirements.

Channel Selectivity and Blocking

IEEE Std 802.15.4 specifies an adjacent channel rejection for service in the 779–787 MHz, 902–928 MHz, 950–956 MHz, and 2.4 GHz bands. In the 868 MHz band, only one channel exists, and an adjacent channel specification is not meaningful. For the 902–928 MHz and 2.4 GHz bands, the receiver must reject an interfering adjacent channel signal that is at the same level (0 dB difference) as a simultaneous on-channel signal. One interfering signal at a time is specified at that level.

In addition, IEEE Std 802.15.4 also specifies an alternate channel rejection for service in the 902–928 MHz and 2.4 GHz bands. The alternate channel is the next nearest channel to the adjacent channel, or two channels away from the channel of operation. The receiver must reject an interfering alternate channel signal that is at a level 30 dB higher than a simultaneous on-channel signal. One interfering signal at a time is specified at that level.

> The specification of these interference levels ensures reliable communication in the presence of multiple colocated WPANs, each on a different channel. The specification of a single interference signal reflects the relatively low traffic envisioned for the service.

To ensure that the receiver does not behave badly in other strong signal conditions, the specification provides a maximum acceptable input level that must not cause an excessive error rate. The IEEE Std 802.15.4 receiver must withstand a signal input of at least –20 dBm without unacceptable error rates. There is no intermodulation specification in any of the IEEE Std 802.15.4 PHYs.

There are four types of service primitives. They are as follows:

- *Request:* The request primitive is passed from the calling layer (also called the user layer) to request that a service is initiated.

- *Indication:* The indication primitive is passed from the service layer to the user layer to indicate an internal event of significance. This event may be logically related to a remote service request, or it may be caused by a service layer internal event.

- *Response:* The response primitive is passed from the user layer to the service layer to complete a procedure previously invoked by an indication primitive.

- *Confirm:* The confirm primitive is passed from the service layer to the user layer to convey the results of one or more associated previous service requests.

PHY SERVICES

The PHY layer provides an interface between the physical radio channel and the MAC sublayer, through the use of two services. These services are the PHY data

> The services of the PHY layer are the capabilities it offers to the MAC sublayer.

service and the PHY management service (called the PHY layer management entity or PLME) and are accessed by the PHY layer data service access point (PD-SAP) and the PHY layer management entity service access point (PLME-SAP), respectively.

PHY Data Services

The PHY data service provides three primitives to the MAC sublayer: PD-DATA.request, PD-DATA.confirm, and PD-DATA.indication, as shown in Figure 4–10. The data service does require a response from the peer MAC and so the response primitive is not required.

PHY Management Services

The PHY management service provides support for commands to control communication settings and the radio control functionality. The PLME primitives are summarized in Figure 4–11. The following paragraphs provide an overview of these primitives.

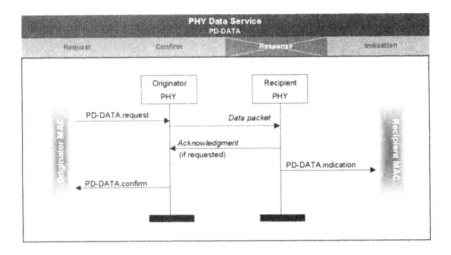

Figure 4–10: Message sequence diagram for data packet exchange mechanism

Primitive	Category	Description	Request	Confirm	Response	Indication
GET	Communication Settings	PHY PAN information base management	X	X		
SET			X	X		
SET-TRX-STATE	Radio Control	Enables/Disables radio system	X	X		
CCA	RF Energy Sensing	RF energy sensing Clear Channel Assessment	X	X		
ED		Energy Detection	X	X		

Figure 4–11: PHY management service primitives

PHY PAN Information Base Management Primitives

The PHY physical layer (PHY)PAN Information Base (PIB) contains configurable attributes to manage the PHY layer. These attributes can be read or written by the PLME-GET and PLME-SET primitives.

Figure 4–12 shows the message sequence diagrams for the procedure of reading or writing the PIB attributes.

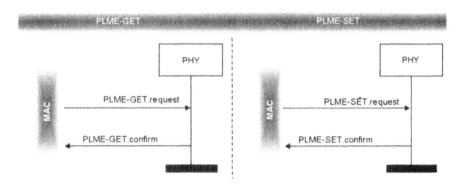

Figure 4–12: Message sequence diagram for PHY PIB reading and writing mechanism

ENABLING AND DISABLING THE PHY

The radio transmitter and receiver can be enabled or disabled by means of the PLME-SET-TRX-STATE primitive. The purpose of this primitive is to control the radio transceiver and enable lower power consumption. Figure 4–13 shows the message sequence diagram for this primitive.

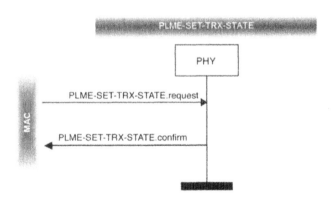

Figure 4–13: Message sequence diagram for transceiver enabling mechanism

CLEAR CHANNEL ASSESSMENT

Before the transmission of packets in a non–beacon-enabled network or in the contention access period of a beacon-enabled network, the MAC instructs the

PHY to perform a clear channel assessment (CCA) before sending data and MAC command frames.

When the clear channel assessment is requested, the PHY layer enables the receiver, performs a CCA measurement, and then disables the receiver. After the clear channel assessment measurement is completed, the PHY layer issues an PLME-CCA.confirm indicating if the channel is busy or not. Figure 4–14 illustrates the CCA mechanism using the PLME-CCA primitives. The figure also shows the message sequence chart for the clear channel assessment mechanism.

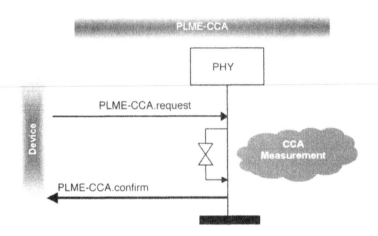

Figure 4–14:   Message sequence diagram for the clear channel assessment mechanism

ENERGY DETECTION

The PLME-ED primitive allows a device to perform RF energy detection in the actual channel where it is operating. The measurement performed is similar to the one carried out in the PLME-CCA primitive, but with higher resolution, it returns an energy level that ranges from 0 to 255. The use of this primitive can enhance the functionality of the network layers. Figure 4–15 shows the message sequence diagram for the energy detection process.

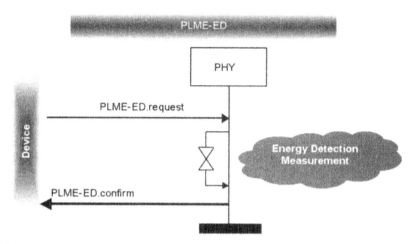

Figure 4–15:   Message sequence diagram for the energy detection mechanism

PACKET STRUCTURE

The PHY Protocol Data Unit (PPDU) is the packet data structure at the PHY protocol layer that modulates the wireless transmitter. The PPDU encapsulates all data structures from higher levels of protocol. The PPDU consists of three components: first, a synchronization header; second, a PHY header; and third, a variable length payload containing the PHY layer service data unit.

PPDU synchronization header: The PPDU synchronization header consists of two fields, a preamble and a start-of-frame delimiter. For all but the optional PSSS and UWB physical layers, the preamble consists of 32 bits, all set to binary zero (recall that a binary zero gets encoded in a chip pattern). For the optional PSSS mode the preamble is 40 bits long when operating at the 868 MHz band and is 30 bits long when operating at the 915 MHz band. In both cases the preamble consists of the first sequence of the respective PSSS code table. The preamble field allows a receiver a sufficient number of bits to achieve chip and bit synchronization. The start-of-frame delimiter consists of the 8-bit pattern 0xe6 (11100101) and allows the receiver to establish the beginning of the packet in the stream of bits.

Note that the start-of-frame delimiter for the optional PSSS mode PHY differs from what is used by the other physical layer options. The start-of-frame delimiter for the PSSS PHY is the inverted first sequence in the respective PSSS code table.

PHY header: The PHY header is a single 8-bit field with the MSB reserved and the remaining low-order bits used to designate frame length information. Packet lengths of 0–4 bytes and 6–8 bytes are reserved. Packets of length 5 bytes are MPDU acknowledgment packets, and packets with 9 bytes or more are MPDU payloads for the MAC protocol layer service.

PHY payload: The PHY payload is composed of only one field called the physical layer service data unit (PSDU). The PSDU is variable length in nature and carries the data payload of the PPDU. All packets carry an MPDU payload for the MAC layer.

Figure 4–16 shows the structure of the PPDU.

PHY Protocol Data Unit

Figure 4–16:   PPDU structure

Chapter 5   Medium Access Control Sublayer

... arbitrating channel access and more

The MAC sublayer, together with the Logical Link Control (LLC) sublayer, comprises the data link layer (also called layer 2) in the ISO/OSI model [29]. The MAC layer provides access control to a shared channel and can offer a reliable data delivery. In the case of WPANs, optimal use of the wireless media is desirable, because they operate in restricted unlicensed bands shared by several other standard and non-standard wireless technologies, including WLANs. IEEE Std 802.15.4 uses a carrier sense multiple access with collision avoidance (CSMA/CA) algorithm, which requires listening to the channel before transmitting to avoid collisions with other on-going transmissions (a kind of wireless etiquette).

The IEEE Std 802.15.4 MAC sublayer has several functions, such as the generation of acknowledgment frames, association, disassociation, security control, beacon generation, and optional guaranteed time slot management, which would be used in a star network topology. The IEEE Std 802.15.4 MAC was designed to allow the implementation of a very thin (simple) protocol stack. This facilitates the quick development of applications and has a direct impact on improving power consumption—the secret is simplicity!

The set of standards from IEEE Std 802 differ at the physical layer and MAC sublayer, but share a common interface at the data link layer. This is achieved through the standardized IEEE 802.2™ LLC. This standard LLC interface was defined before the introduction of wireless networking standards, i.e., WLAN and WPAN. For this reason, IEEE Std 802.15.4 defines the Service Specific Convergence sublayer, which makes the appropriate interface between the MAC and IEEE Std 802.2, while allowing the definition of other LLCs more appropriate for wireless media.

The IEEE Std 802.15.4 MAC sublayer definition contains enhanced functionality normally located in the LLC, making it suitable to be interfaced with the network layer directly, allowing the simple implementation of wireless devices.

The IEEE Std 802.15.4 MAC provides support for two types of wireless network topologies: star topology and peer-to-peer. The management of these types of networks is done in the network layer and is beyond the scope of IEEE Std 802.15.4. In this regard, the MAC performs only functions required by prospective network or other higher layers. Several application scenarios, mostly in the home networking segment, drove the vision for star networks, while for industrial and commercial applications, the interest is more in peer-to-peer networking. The peer-to-peer topology enables the creation of larger ad hoc, self-organizing wireless networks (principally multihop/mesh configurations).

All devices, regardless of topology, participate in the network using their unique 64-bit IEEE addresses; this address can be substituted for a short 16-bit address allocated by the PAN coordinator, thus ensuring that it is unique within its network; this process is managed by the association procedures explained in this chapter.

The 2006 revision of the original standard introduced some new features and modifications to the MAC sublayer. With the exception of security, all these new features are backward compatible. In IEEE Std 802.15.4-2006, the security processing has been simplified and streamlined, reducing the security overhead. Other MAC changes include:

In general, all MAC sublayer changes introduced in the 2006 revision are backward compatible with the original version. However, there are three exceptions: MAC Frames using security processing operations, frames using the channel page field, or frames with MAC payloads larger than 102 bytes are not backward compatible.

- Adding a subfield for indicating the frame version.
- Adding features to facilitate the synchronization of network devices (mechanism to be implemented in higher layer).
- Making use of guaranteed time slots optional.
- Updating primitives to enable using the newly introduced channel pages as a result of the new PHY mode options.
- Scheduling the beacon start time to allow staggering of superframes.
- Simplified broadcasting in beacon-enabled networks.
- Adding features to allow reducing the association time in non-beacon-enabled networks.

STAR TOPOLOGY

In the star topology, communication is controlled by a single PAN coordinator that operates as a network master, sending beacons for device synchronization (including superframe control) and maintaining association management. In this topology, the network devices communicate only with the PAN coordinator, as shown in Figure 5–1.

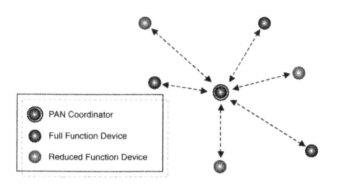

Figure 5–1: Star network topology

Any full function device (FFD) may establish its own network by becoming a PAN coordinator. Each star network operates independently from any neighboring networks. During the formation of a new star network, the PAN coordinator must choose a network identifier, called a PAN ID, that is not used by any other network in its vicinity. This is accomplished by scanning all available or chosen channels for existing networks and then selecting a PAN ID that differs from the ones found. After this procedure is performed, the PAN coordinator can start sending beacons in regular intervals and permitting network devices, requesting to joins its network, to associate.

Network devices programmed to join a star network must first scan for available networks within their radio range by listening for beacons sent by a PAN coordinator. Upon completion of the scan, the higher layer of the network device may select to join one of the networks that were found by sending an association request to the PAN coordinator. The PAN coordinator in turn makes the decision allowing the network device to associate or not.

Star networks also support a non–beacon-enabled mode. In this case, the PAN coordinator uses beacons for association purposes only. Network device synchronization for data exchange is achieved by polling the PAN coordinator for data on a periodic basis.

PEER-TO-PEER TOPOLOGY

The peer-to-peer topology allows any FFD to communicate with any other FFD within its range, and have messages relayed to FFDs outside its range, via multihop routing. This topology enables the formation of more complex, larger networks, including ad hoc, self-organizing, and self-healing structures. IEEE Std 802.15.4 does not specify the details of any of these networks; it only defines the MAC functionality to enable such capabilities.

RFDs may be employed in a peer-to-peer network, but only as peripheral devices, since they lack the capability to relay packets. As a consequence, a sufficient number of FFDs must be present to form the network. Figure 5–2 shows a typical peer-to-peer network.

Peer-to-peer communications requires additional device memory due to the size of the routing tables employed in the higher layer.

A specific type of a peer-to-peer network is the cluster-tree network, consisting of a variable number of coordinators that can serve as a cluster manager or cluster head within a cluster-tree structure but also as a router for relaying messages within the network. The main advantage of a multiclustered wireless network is its hierarchy, which can greatly simplify the routing algorithms used to relay messages across the network. Figure 5–3 illustrates an example of a typical cluster-tree network; the figure shows network devices, cluster heads, and the PAN coordinator that operates as the director of the operation for the entire network.

Similar to the procedure for starting a star network, a new peer-to-peer network is formed by an FFD establishing itself as a PAN coordinator and selecting a PAN ID that differs from other networks within its vicinity. In most cases a peer-to-peer network will be non-beacon enabled, but it can operate in a beacon-enabled mode. Network devices programmed to join a peer-to-peer network must scan for available networks within its vicinity by discovering a nearby PAN coordinator or routing capable devices that act as proxies for PAN coordinators. IEEE Std 802.15.4 refers to routing capable devices as coordinators. Upon completion of the scan, the higher layer of the network device may select to join one of the networks that were found by sending an association request to the PAN coordinator or alternatively to one of the closest coordinators. The coordinator in turn makes the decision allowing the network device to associate or not. If the network device is an FFD, it also becomes a coordinator when joined to a network and provides routing services to other devices in the network or trying to join the network.

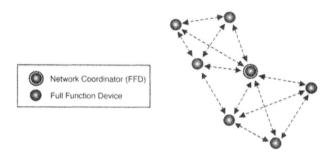

Figure 5–2: Peer-to-peer network topology

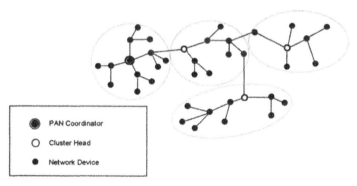

Figure 5–3: Cluster-tree network topology

SUPERFRAME STRUCTURE

IEEE Std 802.15.4 allows the implementation of an optional superframe structure. The superframe is managed by the PAN coordinator and is bounded by beacon messages sent by the coordinator at programmable regular intervals, called the beacon interval. The structure of the superframe can be configured specifically to address the needs of various applications anywhere from small low-latency star networks to large, long-latency, multihop networks.

Each beacon contains information that will help network devices synchronize to the network; this information includes the network identifier, beacon periodicity, and superframe structure. A superframe is divided into 16 contiguous time slots; the first time slot starts at the beginning of a beacon frame.

Network devices that need to communicate with the PAN coordinator must attempt to do so in the time between two successive beacons. This period of time is called the *contention access period* (CAP). To communicate with the PAN coordinator, each network device needs to access the channel using slotted CSMA-CA. Figure 5–4 shows the generic structure of a superframe.

> As a curious historical note, originally the 16 contiguous time slots of a superframe were called "chunks." However, the group found "slot" to be a better term and changed the name accordingly.

Figure 5–4:   Generic superframe structure

> Each GTS is formed by an integer multiple of time slots. Each time slot is equal to 1/16 of the time between the start of two successive beacons.

By request, the PAN coordinator can optionally assign dedicated portions of the superframe to a specific network device. These segments of time are called guaranteed time slots (GTSs). This capability supports applications with a particular bandwidth requirement or that need lower communications latency and is used in star networks. GTSs are all grouped toward the end of the superframe right before the next beacon as illustrated in Figure 5–5. The length of time covering all GTSs is defined as the contention-free period (CFP). Though the CFP can take a significant portion of the superframe, IEEE Std 802.15.4 does require that a minimum amount of CAP space must remain (440 symbols) to allow other devices not using the CFP to access to the channel. The only exceptions to this rule are temporary reductions to accommodate additional beacon lengths for GTS maintenance.

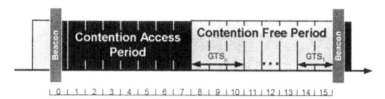

Figure 5–5:   Superframe structure with GTS

The assignment of the GTSs is entirely at the discretion of the PAN coordinator, but the use of contention-free access is useful for applications requiring low latency and dedicated bandwidth allocation.

- *Low latency*: For those applications that require minimum end-to-end delay for prioritized messages, slot assignment facilitates preemptive, out-of-order queue management in the protocols to minimize that delay. Examples include alarm conditions relaying or QoS differentiation based on application software.

- *Bandwidth allocation*: For those applications less sensitive to delay that generate a known data traffic rate, the facility to allocate bandwidth to each service permits the protocol to manage queue lengths at each network device. Traffic management is usually associated with network protocols above the MAC layer, but the capability must be established at the lower levels covered by IEEE Std 802.15.4 to be available to higher levels.

GTSs make higher level protocol services possible, but for those networks that do not implement higher levels of service, the use of GTSs is optional.

For applications with relaxed latency requirements, the superframe may also be split into an active and an inactive portion with the 16 contiguous slots occupying only the active portion of the superframe. An inactive portion of the superframe is not used for communication between devices. Figure 5–6 shows an example where the active and inactive portion are of equal length. However, this is not a requirement and they may be of different proportions. The length of the active portion of the superframe is specified by the super-frame interval, while the frequency of beacons is specified by the beacon interval. This for instance, allows networks implemented with a battery-pow-ered coordinator to reduce the communication duty cycle and increase the bat-tery life of the coordinator.

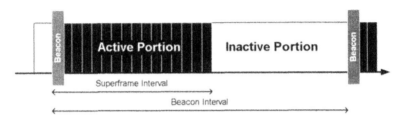

Figure 5–6: Superframe structure with active and inactive portions

A new feature of the IEEE Std 802.15.4 MAC sublayer allows a coordinator to schedule the start of the beacon of a superframe in respect to its coordinator's beacon. Multihop networks can then use the superframe structure, enabling OEMs to create large battery-powered networks.

Another use of this feature is for applications requiring larger beacon-enabled, multihop networks. This feature allows staggering of multiple superframes or without interfering with another. In this case a coordinator (a device providing coordination services to other devices) aligns the start of its active period to fit within the inactive period of its parent's superframe. The parent node may be the PAN coordinator or even another coordinator, as seen in Figure 5–7.

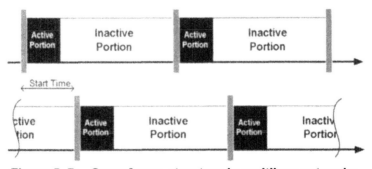

Figure 5–7: Superframe structure in multihop networks

MAC DATA TRANSFER MODEL

The shared nature of the radio-frequency spectrum presents several challenges to the wireless communication system designer. The IEEE Std 802.15.4 MAC was designed to take these challenges into account and provide measures to improve communications reliability by offering a fully acknowledged protocol using CSMA-CA and a frame integrity check.

The data transfer model of IEEE Std 802.15.4 depends on the network topology. In star networks the communication exchange always occurs between a PAN coordinator and a network device, but in the peer-to-peer mode, a device may communicate with its coordinator or any other in its vicinity.

Star networks can have two types of data transfer mechanisms depending on whether the PAN coordinator is beacon-enabled. These are data transfer to a PAN coordinator and the data transfer from a coordinator. A peer-to-peer network supports in addition also the peer-to-peer data transfer.

Data Transfer to a Coordinator

In a beacon-enabled network, a device with data for the coordinator needs to be synchronized to the beacons, sent periodically by the coordinator. If the device is part of a star network and has a GTS assigned, it waits for the appropriate point within the superframe to transmit its data frame without using slotted CSMA. Otherwise the device transmits its data frame in the contention access period of the superframe, in accordance with the slotted CSMA-CA procedure. After receiving the data frame, the coordinator may return an acknowledgment to the network device, if requested; at which point the data transfer is completed. The message sequence diagram for this process is shown in Figure 5–8. IEEE Std 802.15.4 refers to this procedure as *direct data transfer.*

A device with data to send to its coordinator in a non–beacon-enabled network checks the channel availability using CSMA. If the channel is idle the device sends the message to the coordinator. If an acknowledgement is requested, the coordinator confirms the successful receipt of the data frame by returning an acknowledgement to the device. This procedure is outlined in Figure 5–9.

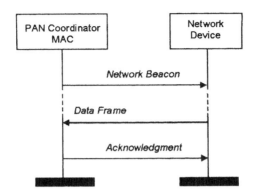

Figure 5–8: Data transfer to a coordinator— beacon-enabled network

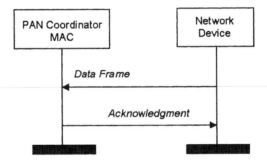

Figure 5–9: Data transfer to a coordinator— non-beacon-enabled network

Data Transfer from a Coordinator

When a coordinator has data to transmit to a device it does not transmit the message immediately but instead appends the device's short address in a special field, called the pending address list, in its beacon. This indicates to the device that data is pending for it at the coordinator. After the appropriate device receives the beacon and detects that data is pending, it sends a data request MAC command frame to the coordinator in the CAP of the superframe. Upon receipt of the command frame, the coordinator responds with an acknowledgment frame followed by the pending data frame. The transaction is completed by an acknowledgment sent from the device to the coordinator indicating the successful receipt of the data frame. If additional messages are pending for the same device the coordinator posts an indication in the beacon of the following superframe. IEEE Std 802.15.4 refers to this procedure as indirect message transfer, the message sequence diagram of this process is summarized in Figure 5–10.

To simplify implementation, the IEEE Std 802.15.4 design team decided not to implement a combined acknowledge/data frame.

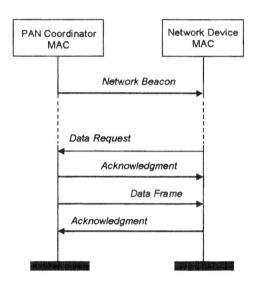

Figure 5–10: Data transfer from a coordinator—beacon-enabled network

Indirect data transfer is also used by a coordinator to send data to a device in a non–beacon-enabled network. However, since there is no mechanism for the coordinator to indicate to the device that a message is pending, the device needs to frequently poll the coordinator for pending messages. The IEEE Std 802.15.4 MAC sublayer has a mechanism allowing the higher layer to poll for messages. Figure 5–11 shows the procedure of the indirect data transfer in a non-beacon- enabled network. When the higher layer of a device instructs the MAC to poll for pending messages, the MAC sends a data request MAC command frame to the coordinator, which confirms the successful receipt of the command with an acknowledgement. This is followed by the pending data frame sent from the coordinator to the device. The transaction is completed by an acknowledgement sent from the device to the coordinator confirming the successful receipt of the data message.

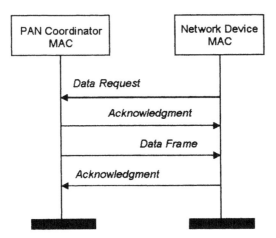

**Figure 5–11: Star network data transfer from PAN coordinator—
non-beacon-enabled network**

Peer-to-Peer Data Transfer

For peer-to-peer topologies, the data transfer strategy is governed by the specific network layer managing the wireless network. A network device might stay in reception mode, scanning the radio-frequency channel for ongoing communications, or might send periodic beacons to achieve synchronization with other potential listening devices.

MAC SERVICES

The MAC sublayer provides two services to the higher layers. These services are the MAC data service and the MAC management service (called the MAC sublayer Management Entity or MLME). They are accessed by the MAC common part sublayer service access point (MCPS-SAP) and the MAC management service access point (MLME-SAP), respectively. For each of these services, IEEE Std 802.15.4 defines a set of protocol primitives that enables the full functionality of LR-WPAN devices. Figure 5–12 shows the detailed IEEE Std 802.15.4 protocol stack architecture and the virtual link between MAC peer entities.

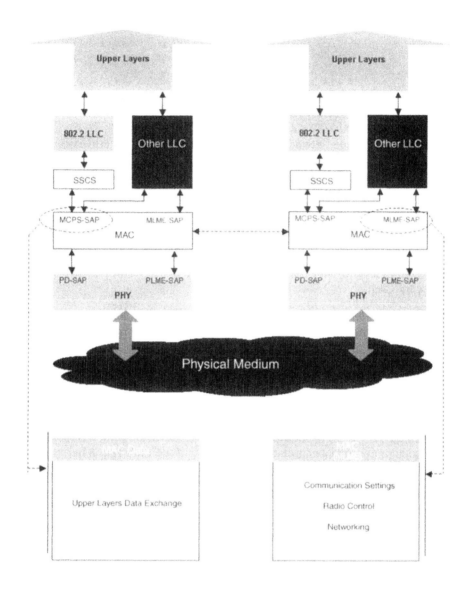

Figure 5–12: IEEE Std 802.15.4 protocol stack architecture

The services of a layer are the capabilities it offers to the next higher layer (or sub-layer). The upper layers build their functionality based on these lower layer services. The MAC services allow the transport of protocol data units between peer entities.

Transmission Scenarios

The packet exchange between two transceivers is susceptible to errors due to the nature of the wireless medium. When one or more errors are present in any given frame, the message will not arrive at the intended MAC recipient because either the recipient PHY layer rejected the packet or the PHY could not synchronize (correlate) with the message itself. In this case, the PHY did not hear the message.

Any acknowledged data transfer has are three possible transmission scenarios, as follows:

- *Successful Data Transmission*: The originator device sends a message to a recipient MAC. The message is received with no errors detected, and an acknowledgment message is sent back to the originator before a timeout occurs. A successful data transmission scenario is illustrated in Figure 5–13.

- *Lost Message Frame*: A message addressed to an intended MAC recipient never arrived at its destination. In this case, the originating device declares a timeout and makes another attempt to send the message. After a number of attempts, the MAC of the originating device sends a notification to its upper layer indicating transmission failure. This transmission scenario is shown in Figure 5–14.

- *Lost Acknowledgment Frame*: The originator of the message transaction did not receive an acknowledgement frame. Similar to the previous case, after a number of attempts, the MAC of the originating device sends a notification to its upper layer indicating transmission failure. This transmission scenario is shown in Figure 5–15.

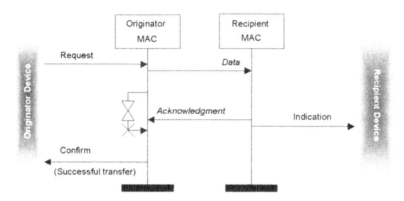

Figure 5–13:   Transmission scenario — successful data transfer

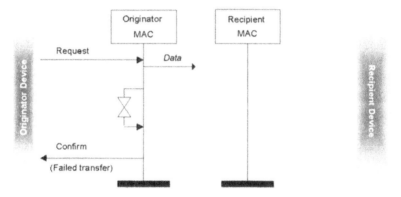

Figure 5–14:   Transmission scenario— lost message frame

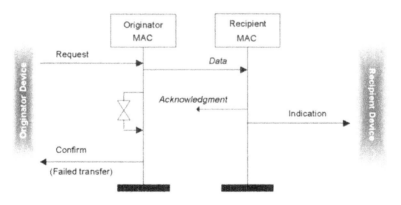

Figure 5–15:   Transmission scenario— lost acknowledgment frame

MAC Data Service

The MAC data service provides three primitives for data transfer: MCPS-DATA.request, MCPS-DATA.confirm, and MCPS-DATA.indication, as shown in Figure 5–16. This figure shows the sequence diagram for a typical data transfer between two devices. The data service does require a response from the peer MAC, and so the response primitive is not required.

The interface offered by the MAC data service has the capability of handling diverse addressing schemes that enable simple implementation of star and peer-to-peer network topologies. In this sense, the MAC enables the upper layers to control which address fields are present in the packet to be transmitted, allowing the implementation of diverse types of network layers, including multihop ad hoc networks. IEEE Std 802.15.4 uses a standard 64-bit IEEE address and short addresses allocated by the association mechanisms. In addition, the MAC data service supports two optional primitives that enable the upper layers to purge an MSDU from the MAC queue. These primitives are MCPS-PURGE.request and MCPS-PURGE.confirm, used for the indirect data transfer. This can be used, for instance, when the information contained in a pending data frame expires before the recipient device gets a chance to retrieve it from the coordinator.

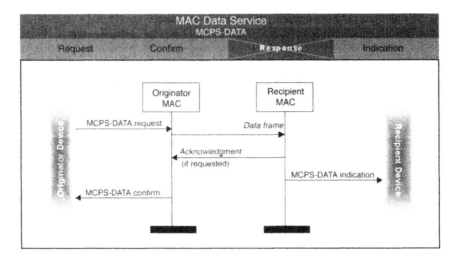

**Figure 5–16: Message sequence diagram for data exchange
mechanism**

MAC Management Service

The MAC management service provides support for commands to control communication settings, the radio control, and networking functionality. The MLME primitives are summarized in Figure 5–17. The following paragraphs provide an overview of these primitives.

To enable the implementation of very low complexity devices, some of the MAC management primitives are optional (i.e., they do not need to be implemented for a product to be IEEE Std 802.15.4 compliant). These primitives are MLME-GTS, MLME-RX-ENABLE, and MLME-SYNC. While other primitives are optional for RFDs only. These primitives are MLME-ASSOCIATE.indication, MLME-ASSOCIATE.response, MLME-ORPHAN, and MLME-START.

Primitive	Category	Description	Request	Confirm	Response	Indication
GET	Communication Settings	MAC PAN information base management	X	X		
SET			X	X		
RESET			X	X		
RX-ENABLE	Radio Control	Enables/Disables radio system	X	X		
SCAN		Scan radio channels	X	X		
ASSOCIATE	Networking	Association control with a network coordinator	X	X	X	X
DISASSOCIATE			X	X		X
GTS		GTS Management	X	X		X
ORPHAN		Orphan device management			X	X
SYNC		Control of device syncronization with network coordinator	X			
SYNC-LOSS						X
START		Beacon Management	X	X		
BEACON-NOTIFY						X
POLL		Beaconless Syncronization	X	X		
COMM-STATUS		Communication Status				X

Figure 5–17:   MAC management service primitives

MAC PAN INFORMATION BASE MANAGEMENT PRIMITIVES

The MAC PAN Information base (PIB) contains configurable attributes to manage the MAC sublayer. These attributes can be read using the MLME-GET primitive and written using the MLME-SET primitive. In addition, the attributes can be reset to their default values with the MLME-Reset primitive.

The MLME-Reset primitive also resets the states of the MAC sublayer and the transceiver.

Figure 5–18 shows the message sequence diagrams for the procedure of reading or writing the PIB attributes.

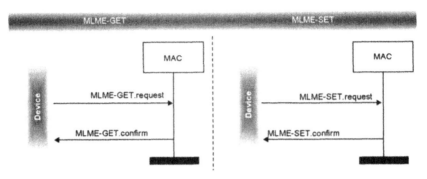

Figure 5–18: Message sequence diagram for reading and writing to the MAC PIB

ENABLING AND DISABLING THE RECEIVER

The radio receiver can be enabled or disabled by means of the MLME-RX-ENABLE primitive. The action can be set for immediate execution or scheduled for a later time. The purpose of scheduling is to enable the MAC to offer network synchronization capabilities to the upper layers.

This primitive allows the power consumption to be lowered in systems that use a network layer that takes advantage of this feature i.e., scheduling data exchange at determined instants, and minimizing power consumption. Figure 5–19 shows the message sequence diagram for this primitive.

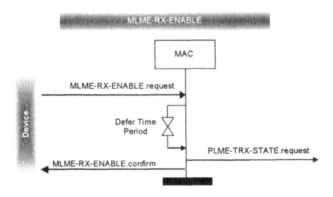

Figure 5–19: Message sequence diagram for enabling the receiver

SCANNING RADIO CHANNELS

MLME-SCAN allows the initiation of a scan over a list of available channels. There are four types of channel scans. They are as follows:

- *Energy Detection Scan*: This scan allows measuring the radio frequency energy in each of the logical channels specified. The measurement is performed by the MLME by issuing a PLME-ED.request primitive and is used by a PAN coordinator to find a suitable channel for starting a new network.

> Beacons are not only sent by PAN coordinators. FFDs associated with a PAN and configured as a coordinator can advertise their presence to other devices by sending beacons for either device synchronization or to facilitate device discovery.

- *Active Channel Scan*: This scan searches for PAN coordinators or coordinators in the radio sphere of influence of devices that are participating in beacon-enabled or non-beacon-enabled networks. For each logical channel, the device first sends a beacon request command (see "MAC Command Frame" section), which causes any PAN coordinator or coordinator to send a beacon. If the PAN coordinator or coordinator are part of a non-beacon-enabled network, they send the beacon using unslotted CSMA. If the PAN coordinator or coordinator are part of a beacon-enabled network, they will send the beacon at the next scheduled beacon interval.

- *Passive Channel Scan*: This scan is used to search for coordinators in the radio sphere of influence of devices, participating in a beacon-enabled network. Passive means that a channel is scanned just by passively listening without sending a Beacon Request Command.

- *Orphan Channel Scan*: Allows an orphaned device (a device that has lost connection with its coordinator) to perform a scan to locate a coordinator. This search is performed over the specified list of logical channels.

The general procedure of the radio scan is shown in the message sequence diagram in Figure 5–20. The scan time is the duration the MAC stays on a particular channel. This time period is based on the smallest superframe size and is calculated using a parameter passed with the primitive from the upper layer.

A device becomes an orphan when it determines that it has lost communication with its coordinator. Communication can be lost for many reasons, including:
- Fading or interference on the existing channel
- A switch of the coordinator to another channel, due to degradation of the existing channel (e.g., interference)
- Motion of either the device or the coordinator, so that the two are now out of range of each other

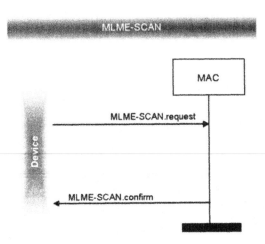

Figure 5–20: Message sequence diagram for the channel scanning mechanism

ASSOCIATION AND DISASSOCIATION CONTROL

A device that is currently not associated to any network uses the channel scan procedure to find prospective candidate networks. If the device wishes to participate in a beacon–enabled network, it uses a passive scan. If it is trying to join a non–beacon-enabled network, it uses an active scan. Following a successful active or passive channel scan the information collected from all the received beacons is passed on to the higher layer. The higher layer may then choose a candidate coordinator to join. If the device is trying to join a beacon–enabled network, it first synchronizes to the periodic beacons of the chosen coordinator. This is done by the use of the MLME-SYNC.request primitive (see "Synchronization Control" section). Once synchronized, the higher layer of the device issues a MLME-ASSOCIATE.request primitive to the MAC sublayer. In a non-beacon-enabled network, the device does not need to synchronize, and the higher layer may issue the association request without synchronizing first.

Upon receipt of an association request primitive, the MAC of the device requesting association sends an association request command to the chosen coordinator. When the command has been received by the coordinator, it returns an acknowledgement to the device. Note that acknowledgements are optional for data frames, but are required for this procedure. The acknowledgment of this message does not imply that the association was accepted; it is simply a confirmation that the command was received. After receiving the association request, the MAC of the PAN coordinator needs to determine if it has sufficient resources to allow another device in its network. The PAN coordinator makes this decision in the time specified in the macResponseWaitTime (a MAC constant parameter), or the requesting device will declare a timeout.

The details of the process for selecting a PAN coordinator after a channel scan are not defined in IEEE Std 802.15.4. The upper layers of a particular implementation must define an association strategy.

Depending on the capabilities and the requirements of the application, the coordinator can accept or reject the association request by issuing an MLME-ASSOCIATE.response with the appropriate parameters. The association response primitive is sent from the coordinator to the device requesting association using indirect data transfer (see "Data Transfer from a Coordinator" section). If coordinator is part of a beacon–enabled network, it indicates in its beacon that a message is pending for the device trying to join. If the device is trying to join a non–beacon-enabled network, it waits for macResponseWaitTime before requesting a response from the coordinator.

Another important feature of the association process is the possibility to request a short 16-bit allocated address from the PAN coordinator. This enables better bandwidth utilization because it reduces the total length of the packet (16-bit as opposed to 64-bit addresses). If the associating device does not request a 16-bit address allocation, the device will participate in the network using its extended 64-bit unique address. Figure 5–21 illustrates the message sequence diagram of the association process.

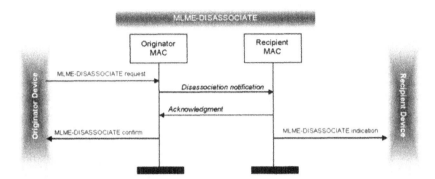

Figure 5–21: Message sequence diagram for association mechanism

A network device or the PAN coordinator can initiate the dissociation process by using the MLME-DISASSOCIATE primitives. The higher layer can request disassociation by issuing the MLME-DISASSOCIATE.request primitive to the MAC sublayer. Then the MAC sublayer sends a disassociation notification command frame using either the data transfer to a coordinator or data transfer from a coordinator, depending on which device initiated the request. The message sequence diagram of this process is shown in Figure 5–22.

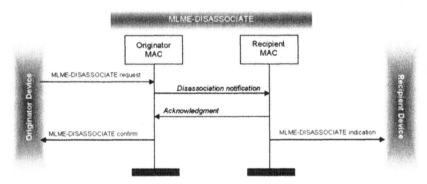

Figure 5–22: Message sequence diagram for the disassociation mechanism

GUARANTEED TIME SLOT MANAGEMENT

As described previously, IEEE Std 802.15.4 supports the optional use of a superframe, which enables the implementation of guaranteed time slots. The

MLME-GTS primitives allow the allocation of new a GTS, deallocation of an existing GTS, or the reallocation of a GTS to eliminate time slot fragmentation. The support for GTS was made optional in the 2006 revision of the standard.

A single GTS can extend over one or more superframe slots. The management of GTSs is performed only by the PAN coordinator, which controls how many of the 16 available time slots are assigned to the contention-free period (the rest are assigned to the contention access period). The PAN coordinator may allocate up to seven GTSs.

> The optional GTS capability allows the implementation of wireless links between a PAN coordinator and a network device with a guaranteed throughput (assuming a reliable link). The PAN coordinator can potentially assign a single GTS covering the whole contention-free period, which in turn may occupy the whole superframe period, less the minimum length contention access period of 440 symbols.

The GTS allocation mechanism is initiated by the network device as indicated in the GTS message sequence diagram of Figure 5–23. The GTS deallocation mechanism can be initiated by the network device or by the PAN coordinator as indicated in the GTS message sequence diagram of Figure 5–24. On reception of a GTS request message, the PAN coordinator determines if space is available within the superframe structure based on the remaining length of the contention period and the number of time slots requested.

When the PAN coordinator is responding to a GTS request, it generates a beacon with its GTS fields indicating the time slot allocated and the number of slots assigned If in the

> Transmissions from network devices in a GTS do not require the use of CSMA-CA.

superframe has no capacity to allocate a new GTS of the desired characteristics, the PAN coordinator will deny the request. In this case, the GTS fields in the beacon indicates the number of time slots remaining in the contention-free period.

> When a network device generates a data request in a GTS-enabled network, the device must defer transmissions until the start of the assigned GTS. This transmission cannot take more time than the allocated time (number of allocated time slots).

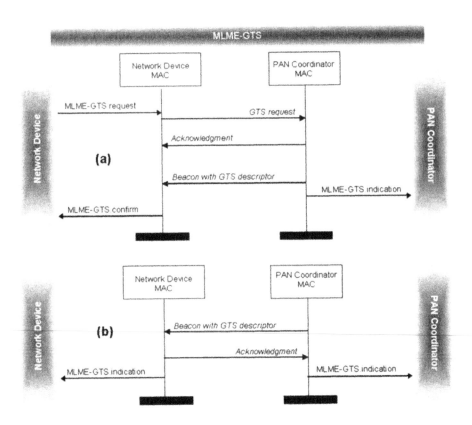

Figure 5–23:   GTS allocation procedure device (a) initiated by the network device and (b) initiated by the PAN coordinator

The GTS message exchange mechanism is fully acknowledged. It concludes with the generation of an MLME-GTS.confirm primitive that includes all relevant information of the GTS allocation procedure. After being assigned a GTS, a network device uses its allocated time to communicate with the PAN coordinator, although the network device can also communicate with the PAN coordinator using the contention access period.

The allocated guaranteed time slots are directional, with data transfer either from the device to the PAN coordinator or from the PAN coordinator to the device.

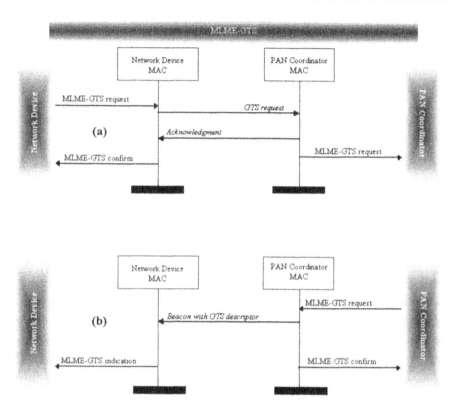

Figure 5–24: GTS deallocation procedure (a) initiated by the network device and (b) initiated by the PAN coordinator

A network device or the PAN coordinator can request the deallocation of an existing GTS. If the operation is initiated by the network device, its MLME should be instructed via an MLME-GTS.request with the appropriate parameters. Similarly, when the PAN coordinator initiates the deallocation of a given GTS, it should send a beacon with its GTS fields indicating that the GTS assigned is being deallocated. The complete deallocation process follows the same steps as the allocation mechanism as presented in Figure 5–24. After the deallocation of a given GTS, the network device can continue communicating with the PAN coordinator using the contention access period.

On a GTS-enabled network, the contention-free period of the superframe can become fragmented as a result of several allocations and deallocations. The PAN coordinator is in charge of removing gaps within the contention-free period in order to ensure optimal use of bandwidth.

If a device loses synchronization with the PAN coordinator, the GTS will be deal-located. After reestablishing a connection with the PAN coordinator (following a scan process), the device can request GTS allocation again.

Orphan Device Management

In the event that a network device loses contact with its PAN coordinator, it performs an orphan channel scan using the MLME-SCAN primitive. As part of the channel scan, the MAC of the network device sends orphan notification command messages in each of the available and specified channels. When the MAC of the PAN coordinator or a coordinator receives this notification, it generates an MLME-ORPHAN.indication, which causes the coordinator to verify whether the network device was previously associated with its net-work. If the network device was associated to the coordinator, it generates an MLME-ORPHAN.response. At this point, the MAC of the coordinator will send a coordinator realignment command. In the event that the network device was not previously associated with the PAN coordinator, there will be no response from it. Figure 5–25 shows a message sequence diagram for the orphan notification mechanism.

A reduced function device is not required to implement the MLME-ORPHAN primitives.

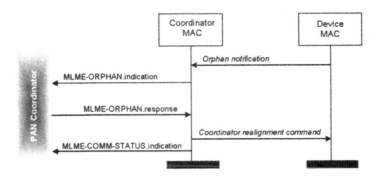

Figure 5–25: Message sequence diagram for orphan notification mechanism

SYNCHRONIZATION CONTROL

Synchronization control with a coordinator is achieved with the use of the MLME-SYNC and MLME-SYNC-LOSS primitives. MLME-SYNC allows the network device to locate and track beacons in a beacon-enabled PAN. The search process is initiated with an MLME-SYNC.request. The search is done by activating the radio receiver and waiting, for a given amount of time, for a beacon frame to arrive from a coordinator.

In a beacon-enabled PAN, the search for beacons can be performed in one of two modes: one mode in which the MAC of the network device continuously tracks its coordinator's beacons; or another mode in which the MAC locates the beacon only once. In both cases, if a beacon is received indicating data pending for the receiving network device, it sends a data request command to the PAN coordinator. The message sequence chart for network device synchronization is shown in Figure 5–26. The MLME-SYNC.request is not used in non–beacon-enabled PANs.

In the event of a loss of synchronization with the coordinator, the MAC generates an MLME-SYNC-LOSS.indication. Four possible situations can cause a loss of synchronization event. They are as follows:

- *Beacon Lost*: After MLME-SYNC, either initially or during tracking, the beacon message has not been received. The message sequence chart for this scenario is shown in Figure 5–27.

- *Coordinator Lost*: Several attempts to communicate with the coordinator have failed.

- *PAN ID Conflict*: The network device detected a PAN ID conflict, defined as two different PAN coordinators with the same ID being in range of the network device.

- *Realignment*: The network device received a coordinator realignment command message from the PAN coordinator.

BEACON MANAGEMENT

Beacon generation is initiated with the use of the MLME-START primitive. The parameters of this primitive allow configuring the device as a PAN coordinator or a coordinator, selecting a logical channel, establishing beacon periodicity, and setting up the superframe characteristics. After the MLME-

START.request, the MAC will respond with an MLME-START.confirm, as illustrated in Figure 5–28. When a network device receives a beacon frame containing the PAN ID of the network with which it is associated, the MAC interprets its content. If the beacon frame contains one or more octets of payload (data), the MAC issues an MLME-BEACON-NOTIFY.indication event. Figure 5–29 illustrates the beacon notification mechanism.

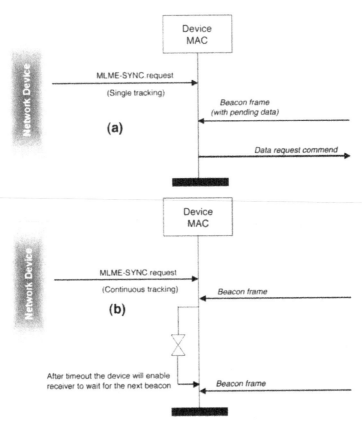

Figure 5–26: Message sequence diagram for network device synchronization (a) single tracking and (b) continuous tracking

Full-function devices operating in a peer-to-peer network can generate beacon messages to neighboring devices. These beacons can act as Hello messages during network formation. If the MLME-START is issued with the *PANCoordinator* parameter set to FALSE, the device generating beacons is called a coordinator. In IEEE Std 802.15.4, only full-function devices can become coordinators and have the capability of generating beacons.

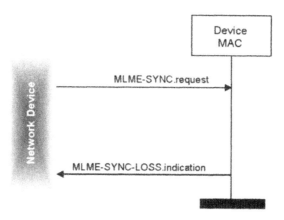

Figure 5–27: **Message sequence chart for loss of synchronization due to beacon lost scenario**

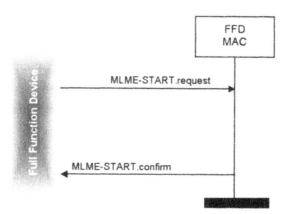

Figure 5–28: **Message sequence chart for procedure to start generating beacon transmissions**

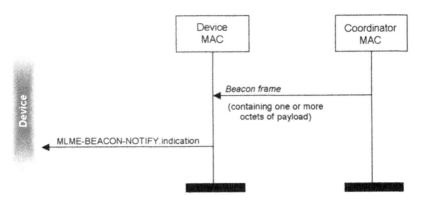

Figure 5–29: **Message sequence chart for beacon notification mechanism**

BEACONLESS SYNCHRONIZATION

For non-beacon-enabled networks, a network device can use the MLME-POLL primitive to poll a coordinator for pending data at its discretion. When the MAC sublayer of a network device receives a MLME-POLL.request, it sends a data request command frame to its coordinator. The frame pending flag (FP) in the frame control field of the acknowledgement following the data request indicates whether data is pending. After a MLME-POLL.request, the MAC will generate an MLME-POLL.confirm with the results of the polling procedure. Figure 5–30 shows the message sequence chart for this primitive

If the network device receives a beacon generated by its coordinator, the MAC checks its device address in the address list field from the beacon frame. If its address is listed, the MAC generates a *data request* message to the coordinator, as indicated in the data transfer from the coordinator (beacon enabled).

COMMUNICATION STATUS

With this primitive, the MAC generates a variety of communication status messages to its upper layer. The MLME-COMM-STATUS.indication is invoked when any of the following conditions occur:

- After a transmission caused by a response primitive
- After reception of a frame that did not pass the security procedure

The different amendments to the IEEE Std 802.15.4-2006 revision have caused the addition of MAC functionality enhancements. The enhanced functionality was added to support PHY specific constraints to comply with regional regulatory conditions, e.g., operation in the 950 MHz Japanese band or additional functionality to support the location awareness capability offered by the IEEE Std 802.15.4a amendment.

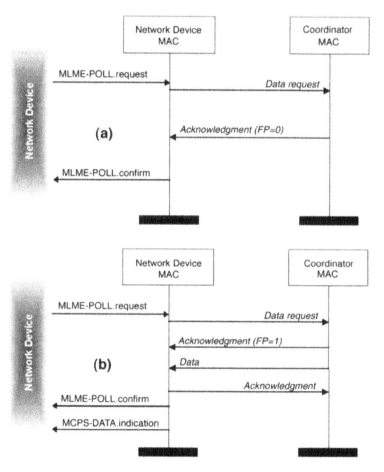

Figure 5–30:   Message sequence chart for polling the coordinator (a) no data pending and (b) with data pending

MAC FRAME STRUCTURE

The IEEE Std 802.15.4 MAC frame structure is designed in a way that reflects the protocol simplicity and flexibility while having the minimal elements to overcome the challenges of the wireless media. A MAC frame consists of three parts: header, variable length payload, and footer.

The MAC header contains a frame control field, a sequence number field, an addressing field, and optionally an auxiliary security header field. The frame control field specifies the type of frame, security usage, and the format and content of the address fields. The frame control field also indicates if an acknowledgment from the recipient of the frame is required. The sequence number field contains a number that is increasing with each transmitted frame. The address field contains the source or destination addresses as specified in the frame control field. If security is used (as indicated by a subfield in the frame control field) the frame may also include an auxiliary security header field. Its length may vary and depends on the specified security level.

Depending on the type of MAC frame and the network topology in which a device is operating, a MAC frame can contain one or more of the following: the source (originator) PAN ID, source device address, destination PAN ID, and destination address. With this feature, the application can use diverse network topologies.

The MAC payload contains information specific to the type of transaction being handled by the MAC and can be logically divided in several fields for use by upper protocol layers.

Finally, the MAC footer consists of a 16-bit frame check sequence (FCS) based on the standardized ITU-T 16-bit cyclic redundancy check (CRC) algorithm (formally called CCITT 16-bit CRC). The general MAC frame structure is shown in Figure 5–31.

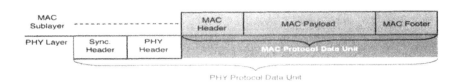

Figure 5–31: General MAC frame structure

When the three components of the MAC frame are assembled into the PHY packet, it is called the MAC protocol data unit (MPDU).

IEEE Std 802.15.4 defines four types of MAC frames: Beacon, Data, Acknowledgment, and MAC command. These frames are described in the following paragraphs.

Beacon Frames

In a beacon-enabled network, a full-function device may transmit beacon frames. In a beacon frame, the address field contains the source PAN ID and the source device address. The MAC payload of a beacon frame is divided into four fields:

- *Superframe Specification Field*: Contains the parameters that specify the superframe structure (if any).

- *Pending Address Specification Field*: Contains the number and type of addresses listed in the Address List Field.

- *Address List Field*: Contains a list of device addresses with data available at the PAN coordinator.

- *Beacon Payload Field*: Optional field that may be used by higher layers. For example, it could be used to broadcast data to the devices participating in its network within its range of coverage.

The format of the beacon frame is shown in Figure 5–32.

MAC Header				MAC Payload					MAC Footer
Frame Control	Sequence Number	Addressing Fields	Auxiliary Security Header	Superframe Specification	GTS fields	Pending address field	Beacon Payload		FCS

Figure 5–32:   Beacon frame format

Data Frames

Data frames are used by the MAC sublayer to transmit data. The address field contains the PAN ID and device ID of the source and/or destination, as specified in the MCPS-DATA.request primitive. The format of the data frame is shown in Figure 5–33.

MAC Header				MAC Payload	MAC Footer
Frame Control	Sequence Number	Addressing Fields	Auxiliary Security Header	Data Payload	FCS

Figure 5–33: Data frame format

Acknowledgment Frame

Acknowledgment frames are sent by the MAC sublayer to confirm successful frame reception to the originator of a message. The acknowledgment frame is generated only if the received message is requesting acknowledgment and the FCS (frame check sequence) is evaluated as good by the receiving device.

The acknowledgment frame does not contain the address field in the MAC header. Similarly, there is no MAC payload.

The acknowledgment was designed to be very short to minimize network traffic. Recall that when a message acknowledgment is required, there will be one acknowledgment per MAC frame.

When a network device receives an acknowledge frame, it first verifies that it was expecting one and then matches the received frame sequence number with the one it is expecting; otherwise, the acknowledgment frame is discarded. Figure 5–34 shows the format of the acknowledgment frame.

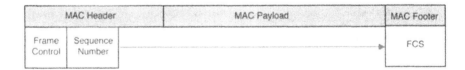

MAC Header		MAC Payload	MAC Footer
Frame Control	Sequence Number		FCS

Figure 5–34: Acknowledgment frame format

MAC Command Frame

The MAC command frame is originated by the MAC sublayer and is in charge of all the MAC control transfers for each MAC command type shown in Table 5–1.

Table 5–1:   MAC command frame types

Command Identifier	Command Type
1	Association Request
2	Association Response
3	Disassociation Notification
4	Data Request
5	PAN ID Conflict Notification
6	Orphan Notification
7	Beacon Request
8	Coordinator Realignment
9	GTS Request
10-255	Reserved

The MAC payload has two fields, the MAC command type and the MAC command payload. The MAC command payload contains information specific to the type of command in use. Figure 5–35 shows the details of the MAC command format.

MAC Header				MAC Payload		MAC Footer
Frame Control	Sequence Number	Addressing Fields	Auxiliary Security Header	Command Type	MAC Command Payload	FCS

Figure 5–35:   MAC command frame format

MAC FUNCTIONAL SCENARIOS

IEEE Std 802.15.4 contains a detailed explanation of the mechanisms that define the MAC functionality. The following paragraphs provide a brief overview of these mechanisms.

Accessing the Channel

In IEEE Std 802.15.4, the physical radio channel is accessed using CSMA-CA before attempting to send any frame. Exceptions to this are beacon frames, GTS transmissions, acknowledgment frames, and data frames following a data request command frame if the MAC sublayer can return the data frame within 12–32 symbols after the command frame.

In beacon-enabled PANs, slotted CSMA-CA is used, whereas in non-beacon-enabled PANs, the basic unslotted CSMA-CA algorithm is implemented.

Starting and Maintaining PANs

Network devices attempting to participate in a network must proceed to locate it by scanning the RF channels in its list of available channels. As previously described, the MLME-SCAN primitive allows performing active and passive scanning for beacons. After a beacon of a suitable coordinator is found, the network device initiates the association process.

Devices that lose communications with a PAN coordinator or their coordinator perform an orphan scan through its list of available channels. Auy.p the coordinator receives the inquiry from the orphaned device, it sends back a coordinator realignment command.

After a channel scan, if there are no appropriate networks to join, an FFD may begin operating as a PAN coordinator (sending beacons) through the use of the MLME-START.request primitive. In some instances, it is possible that two PAN coordinators with the same PAN ID are collocated. One reason for this problem is that a PAN coordinator self-assigns its PAN ID without consulting with neighboring coordinators.

A PAN coordinator detects a PAN ID conflict if it receives a beacon with its own PAN ID from a coordinator acting as a PAN coordinator or when a PAN ID conflict notification command is received from one of its child devices. Similarly, a network device can detect a conflict when it receives a beacon with the same PAN ID of the network it belongs to, in a time that has no correlation with the beacon period or when this beacon carries a source device address different from the address of its original PAN coordinator. In either case, the network device sends a PAN ID conflict notification command to its PAN coordinator.

After the PAN coordinator detects the conflict, it performs an active channel scan that enables the selection of a new PAN ID. Then it broadcasts a coordinator realignment command to its associated network devices.

The procedure for a PAN coordinator to assign its PAN ID is outside the scope of IEEE Std 802.15.4.

Device Synchronization

In a beacon-enabled network, devices are required to synchronize with the beacon frames to detect any pending messages. In a non–beacon-enabled network, devices also send beacons acting as "hello" messages to enable discovery by neighboring devices.

GTS Management

The PAN coordinator is in charge of maintaining the complete administration of the superframe structure and the control of the allocation, deallocation, and reallocation of guaranteed time slots.

Network devices can request GTSs that extend over several superframe time slots. The PAN coordinator allows the request, depending on the availability of time slots in the contention-free period and the needs of the overall network. The MCPS-DATA.request primitive contains a parameter that indicates if the frame will be transmitted using a GTS or if it will be transmitted using the contention access period. When a GTS transmission is requested, the transmission is deferred until the start of the assigned GTS.

A network device or the PAN coordinator can initiate GTS deallocation. A network device uses the GTS request command, while the PAN coordinator indicates any changes to the GTS allocation in its beacon. A PAN coordinator initiates GTS reallocation after detecting GTS fragmentation due to previous GTS deallocations.

SECURITY SERVICES

The MAC of IEEE Std 802.15.4 provides security services that are controlled by the MAC PIB. The MAC sublayer offers two security modes: an unsecured mode and a secured mode.

To meet the objectives associated with the security modes, a critical function of IEEE Std 802.15.4 MAC is frame security. Frame security is actually a set of optional services that may be provided by the MAC to the upper layers. The standard strikes a balance between the need for these services in many applications and minimizing the burden of their implementation on those applications that do not need them. The available services are described in the following sections.

Due to the variety of applications targeted by IEEE Std 802.15.4, the processes of authentication and key exchange are not defined in the standard.

Data Confidentiality

Data confidentiality is provided through encryption using a symmetric cipher, in which the same key is used to encrypt plaintext at the message source and then decrypt the resulting ciphertext at the message destination. Devices without the key cannot decrypt the message. The standard defines encryption of beacon payloads, command payloads, and data payloads. Other messages, such as acknowledgments, and message components, such as addresses, are not encrypted.

Data Authenticity

Data authenticity, also called data integrity, is a service that enables a receiving device to detect the modification of a message by parties without the correct cryptographic key, by appending a message integrity code (MIC) to the message. The standard defines integrity checking of MAC header, auxiliary security header, and unsecured payload fields of data, beacon, and MAC command frames. Integrity is expected to be the security service most used by IEEE Std 802.15.4 applications. Unlike file transfer or voice communication applications common with other protocols, IEEE Std 802.15.4 applications often transmit messages that do not convey secret information. What is often more important is that the messages are authenticated (i.e., the source is known and trusted), and that any tampering with or modification of the message is detected. For example, the fact that a light in a building is turned off is usually not secret; however, it is more important to be sure that any message to the light instructing it to turn off comes from a trusted source (e.g., the room switch), rather than a prankster or cracker.

To avoid replay attacks to an IEEE Std 802.15.4 network, in which an old message is stored by a malicious entity without the cryptographic key and then replayed later, the data authentication service places one of an ordered sequence of values in the auxiliary security header. When received, this value is compared against a stored value; if it is newer than the stored value, the replay protection check passes and the new value is stored. Although the replay protection can determine that one message is newer than another, it is a relative determination; no statement of absolute time is made. The use of the data authentication service requires additional memory to store the present value, and increased latency to transmit the message integrity code.

IEEE Std 802.15.4 provides combinations of these three services in two security modes, which are designed to serve a wide range of applications.

> The security service of IEEE Std 802.15.4 has been simplified and streamlined in the 2006 revision. The result is that the secured mode is not backward compatible. However, communication between devices following the 2003 and 2006 editions is still possible by using the unsecured mode.

Unsecured Mode

In unsecured mode, no security services are provided. This mode is suitable for some applications in which implementation cost is important, and security is either not required or obtained in other ways. Examples of this type of application include an advertising kiosk in a public place, or a low-power wireless motor controller in the center of a large, physically secure area outside of which its signal cannot be detected.

Secured Mode

In secured mode, the device may offer two security services (data confidentiality and data authenticity), depending on the security level employed. Eight security levels have been defined as shown in Table 5–2. The unsecured mode is defined as level 0. This level provides the backward compatibility with deployed devices following the 2003 edition of the standard.

> When a secured mode with data authenticity is used, replay protection is always provided.

Table 5-2: IEEE Std 802.15.4 security levels

Security Level	Security Attributes	Data Confidentiality	Data Authenticity	Replay Protection
0	None	OFF	NO	NO
1	MIC-32	OFF	MIC-32	YES
2	MIC-64	OFF	MIC-64	YES
3	MIC-128	OFF	MIC-128	YES
4	ENC	ON	NO	NO
5	ENC-MIC-32	ON	MIC-32	YES
6	ENC-MIC-64	ON	MIC-64	YES
7	ENC-MIC-128	ON	MIC-128	YES

The AES-128 (Advanced Encryption Standard, with 128-bit keys and 128-bit block size) symmetric-key cryptography algorithm [27] is employed for security levels 1 to 7.

Security levels one to three provide the data authentication service with integrity codes of 32, 64, or 128 bits in length to provide data authenticity and replay protection. Security level four provides data confidentiality service, while security levels five to seven employ AES to provide data confidentiality and data authenticity (with message integrity codes of either 32, 64, or 128 bits). The length of the integrity code does not refer to the strength of the AES algorithm, but rather to the number of bits of the code actually transmitted with the message.

This set of security suites offers several advantages. A major advantage is that it employs only one cryptographic algorithm for all levels of security. Other alternatives require additional algorithms, (for example, a hash algorithm for integrity checking), to provide a complete multilevel security solution. This results in a larger and, therefore, more costly implementation. In particular, by reusing AES in a clever way, the AES-CCM* modes enable a single algorithm to provide both security services in a very small implementation. The CCM* is an extension of the counter with cipher block chaining message authentication code mode of operation. The use of AES-CCM* modes retains compatibility with other IEEE 802 standards, such as IEEE Draft Std 802.11i and IEEE Std 802.15.3. This not only allows circuit and code reuse between the standards; it ensures that a larger audience is reviewing and studying the security algorithms for weaknesses. Over time, this high level of interest and analysis increases confidence in the security algorithms.

Chapter 6   Network Functionality

...getting information from source to destination

IEEE Std 802.15.4 was designed to support multiple network topologies—from a star network to many network types based on peer-to-peer communication, including mesh, tree, cluster, and cluster-tree networks. Although the network layer per se is beyond the scope of the standard, this chapter describes some sample networks that can be constructed using IEEE Std 802.15.4, including samples of their associated message–routing algorithms. This chapter also provides an overview on message routing algorithms to illustrate the type of performance that IEEE Std 802.15.4 networks can achieve and the wide variety of applications to which they can be applied.

FEATURE OVERVIEW

IEEE Std 802.15.4 supports star networks in a master–slave configuration and a general form of peer-to-peer communication, which can be used to construct a variety of network types. The star network (Figure 6–1) is a single- or double-hop network (i.e., all network devices are in range of a single common device), whereas peer-to-peer networks, for example, mesh (Figure 6–2) or cluster (Figure 6–3) networks, may be multihop networks (i.e., the message source and destination devices are not necessarily within range of each other, and communication may occur through multiple intermediate devices that relay messages). The standard thus supports both single-hop and multihop communication links.

Starting a Network

As described in Chapter 5, an IEEE Std 802.15.4 network is started by an FFD when it becomes the PAN coordinator. All networks must have exactly one PAN coordinator in each network. Figure 6–4 shows the network formation, which starts when an upper layer of the FFD sends the MLME-

SCAN.request primitive to the MLME, requesting an active channel scan. Upon completion of the scan, the results are sent back to the upper layers by the MLME-SCAN.confirm primitive. If the results of the scan are acceptable, the upper layer selects a PAN ID (which may be predetermined) and sends the MLME-START.request primitive (with the PAN coordinator parameter set to TRUE) to the MLME.

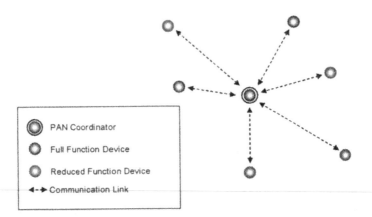

Figure 6–1: Star network

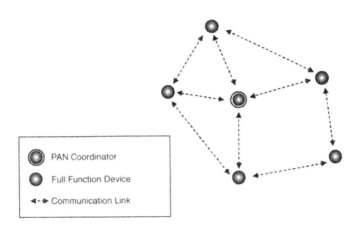

Figure 6–2: Mesh network

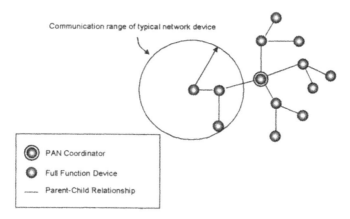

Figure 6–3: Cluster network

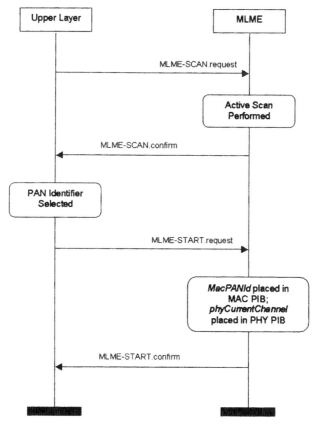

Figure 6–4: Network formation

The MLME-START.request primitive requires the MAC sublayer to place the PAN ID macPANId value in the MAC PIB and the logical channel phyCurrentChannel value in the PHY PIB. After this process is completed, the MAC sublayer sends the MLME-START.confirm primitive to the upper layers, the FFD begins operating as a PAN coordinator, and the network is started.

All IEEE Std 802.15.4 devices have a unique 64-bit IEEE address, aExtendedAddress (often called a MAC or physical address), stored in the MAC PIB table. When a device joins a network, it identifies itself by its aExtendedAddress value; at the discretion of the PAN coordinator, it can exchange that address for a shorter, network-specific logical address supplied by the PAN coordinator during the network association process.

UPPER LAYER NETWORK FORMATION POLICIES AND ALGORITHMS

As previously indicated in Chapter 5, some policies and algorithms associated with network formation and association are performed by upper layers of the protocol and not defined in the standard. A few alternatives for these algorithms are described here. Recall that these alternatives are only possibilities—other network-forming policies and algorithms could be used as well.

PAN Coordinator Selection

The first step in network formation is the selection of the PAN coordinator. The policy deciding the transmission of the MLME-START.request primitive (with the PAN coordinator parameter set to TRUE) from an upper layer to the MAC is, of course, determined by that upper layer. One can envision several application-dependent scenarios as follows:

- *Dedicated PAN coordinator*: There are some applications, for example, some home security systems, for which it is clear that only one device, the gateway device, needs to be connected to an outside network. This device should be the PAN coordinator. These applications require the consumer to purchase the entire system as a whole at one time, so that the manufacturer has complete control over the final network design and behavior. In other situations, the consumer may have the flexibility to purchase network devices individually. This policy may be reinforced by employing only one FFD in the network (as the PAN coordinator) and populating the rest of the network with RFDs. In such a network, exactly one device is eligible to become the PAN coordinator.

- *Event-determined PAN coordinator*: Other applications might have a large number of identical devices, any one of which could become the PAN coordinator with an external stimulus. (For example, a user presses a button.) In this case, every device must be an FFD.

- *Self-determined PAN coordinator*: For these applications, the goal is to establish a network, but it is less important which device in the network is the PAN coordinator. An example of this type of application is a location-determining network. The purpose of this network is to determine the relative location of each network device, possibly by means of a distributed algorithm. This type of application might not have an external gateway; so any network device could serve as the PAN coordinator. One way of forming this type of network is to employ a form of power-on-network formation, in which the upper layer of each device instructs the MAC to begin an active channel scan (the step before network formation) when a device starts up. The first device to complete this scan (which should produce a negative result if the network is not yet formed) and send the result to the upper layer may be rewarded with the MLME-START.request primitive and become the PAN coordinator. As always, all prospective PAN coordinators must be FFDs.

Selection of the PAN ID

After the channel scan has been completed and the results returned to the higher layers, the results should be evaluated and action taken by the higher layers.

In some applications, it may seem desirable from a security standpoint to fix the desired PAN identifier ahead of time so that new network devices will be limited to joining only one particular network (that must be already established). Although possible with IEEE Std 802.15.4 networks, this policy is discouraged, because of the relatively small number of PAN IDs available and the unpredictable behavior (from the user's perspective) that could result if two networks with identical PAN IDs were co-located. (There is a procedure for PAN ID conflict resolution in IEEE Std 802.15.4.) An alternative method is to require all network devices upon association to check the 64-bit extended address of the PAN coordinator and compare it with the desired address. In this way, association with the correct network, regardless of PAN ID, can still be guaranteed.

If a new PAN will be created in the same region as a preexisting PAN (a case in which a preexisting or newly placed FFD becomes a dedicated PAN coordinator for the new PAN), the new PAN coordinator must posses a different PAN ID than all PAN coordinators it is capable of hearing. However, if the goal is for an FFD to become a member of any network, and assume PAN coordinator status only if no PAN IDs were received during the channel scan, many scenarios are possible.

If exactly one PAN ID has been received during the channel scan, the device might attempt to associate with the device transmitting that PAN ID. If more than one device was heard transmitting a PAN ID, some type of selection algorithm must be employed to determine to which device association should be attempted. This algorithm may consider other information transmitted in the received beacons, consider previous history (perhaps previous association attempts with a particular device have failed), or may employ a simple deterministic or stochastic process (e.g., round-robin, first-in, or random selection).

The final scenario, of course, is the case in which no other PAN IDs were heard and, for application reasons, the device is not permitted to become a PAN coordinator. In this scenario, the device might return to sleep for an application-dependent period, waking later to repeat the active channel scan.

The Use of Beacons

Beacon use is required in IEEE Std 802.15.4 networks for network discovery (i.e., when the network wishes to attract new network devices). Beacons also are useful for coordination and synchronization of an active network (e.g., for GTS control), and in applications for which it is desirable to minimize coordinator-to-network device message latency, because the network device can readily check the message pending field of a beacon to determine when messages are waiting for it.

However, beacons can be disadvantageous in some applications. For example, if no messages are expected from the coordinator to the network device and only light traffic is expected from the network device to the coordinator (for example, in a wireless light switch), beacon transmission will waste the power of the coordinator, and reception of the beacons will waste the power of the network device. A better solution is to stop the beacon transmissions and have the network device asynchronously transmit its (rare) data frames as they are generated. This approach is especially effective if the network is in a star configuration, and the PAN coordinator is externally powered. In this case, the PAN coordinator may operate constantly in receive mode, while the (presumably battery-powered) network device need expend significant energy only when it transmits a state change to the PAN coordinator. As a result, the battery life of the network device can be greatly extended. This is true even if occasional coordinator-to-network device traffic is expected, requiring the network device to poll the coordinator for data.

THE STAR NETWORK

A star network is a good design choice for applications that cover a limited physical area, so that a single device (the master, which will be the PAN coordinator) is in range of all other network devices (the slaves). IEEE Std 802.15.4 star networks must employ an FFD as the PAN coordinator; however, the other network devices may be either FFDs or RFDs. Because star network devices communicate only with the PAN coordinator, they have potential implementation cost savings over peer-to-peer networks, which must store information for every peer with which they communicate. An IEEE Std 802.15.4 RFD, operating as a network device in a star network, can be a good alternative for simple point-to--point applications that require an extremely low-cost implementation.

Because star networks by their nature are single- and double-hop networks, message latency can be lower than in multihop networks. If latency is a critical performance metric of the desired application, a GTS may be assigned by the PAN coordinator to reserve time for a particular network device in each superframe, avoiding the delay of contention-based channel access and guaranteeing a fixed amount of bandwidth to the network device. In this way, an IEEE Std 802.15.4 star network can provide maximum message latencies as low as 15.36 ms, suitable for use in PC peripheral applications such as wireless mice and joysticks. In the extreme case, it is possible to extend a single GTS to encompass all time between beacons. A single network device can then have the entire channel bandwidth, in excess of 115.2 kb/s for the 2.4 GHz band, for relatively high-bandwidth applications.

Conceptually, message routing in star networks is viewed differently than message routing in peer-to-peer networks. Because the PAN coordinator of a star network can hear all network devices and directly controls access to the shared channel, routing in star networks is viewed as occurring in the MAC layer, as packet switching, rather than in the network layer as part of a peer-to-peer message routing algorithm. With this idea in mind, message routing in star networks has been defined in IEEE Std 802.15.4. In fact, routing at this level can be shown to have some implementation advantages, because simple message relaying need not involve the higher layers of the protocol.

PEER-TO-PEER NETWORKS

The peer-to-peer communication capability of IEEE Std 802.15.4 devices allow the creation of many types of peer-to-peer networks, each with their own advantages and disadvantages. A sampling of network types possible with IEEE Std 802.15.4 FFDs is described in the following sections.

The Flat Mesh Network Topology

The simplest type of peer-to-peer ad hoc network to envision is perhaps the flat mesh network, a network composed of a number of identical network devices, not all of which may be in range of any one device. Messages may be relayed from source devices to destination devices by a large number of routing algorithms.

Although the network devices in the mesh may be identical, one must have special capabilities to perform the PAN coordinator function, give the network its PAN ID, and control the association of new devices to the network. However, the PAN coordinator need not take a role in message routing.

A true mesh network has regularly spaced devices forming a repeating grid pattern of communication links; however, the definition used here, allowing an ad hoc device distribution, has become common.

Routing

A principal problem to be overcome in a flat mesh network design is the problem of addressing. Because the network is logically flat (i.e., there is no hierarchy) and there is no other grouping or organization of network devices, the address of a network device does not provide clues to the route needed to get a message to it. Nevertheless, many routing algorithms have been devised for such networks, a few of which are listed here:

- *Flooding*: The simplest method to route messages is to send a copy of each message to all network devices. Although this is a correct algorithm (i.e., it will in fact deliver the message to the intended recipient), it is not very efficient, especially for the large networks IEEE Std 802.15.4 can support. Much power can be wasted sending the message to unintended recipients. This method is seldom used in practical networks of any size to transmit data traffic, although it is needed in nearly all networks for at least some control and status message traffic.

- *The Bellman-Ford algorithm*: The Bellman-Ford algorithm requires all network devices to maintain a routing table that contains routing costs (usually the number of hops, although more complex cost accounting, such as available power of network devices in the route, can be used) of the optimum route to all other network devices, plus the address of the first device in that route. Devices maintain their tables by exchanging them with all devices within their range, and then comparing entries by destination. The route of a message is determined by the source device in advance (*source routing*) and placed in totality inside the MSDU, as part of the message payload. Relaying network devices follow these routing instructions to deliver the message. Although an elegant solution to the routing problem, the Bellman-Ford algorithm suffers from poor dynamic behavior. If a communication link is broken, for example, its convergence behavior to the new set of optimal routes is not good. In addition, because the routing tables must have as many entries as there are network devices in the network, the costs to exchange them become prohibitive once the network becomes large.

- *The GRAd (Gradient Routing for Ad hoc networks) algorithm*: The GRAd algorithm [21] requires all network devices to maintain a cost table that lists the routing cost to each potential destination device (as opposed to the Bellman-Ford algorithm, which lists all network devices). However, network devices do not exchange cost tables, and messages are not sent by unicast transmission to particular network devices along a route determined by the message source. Instead, messages are broadcast by the source to all network devices in range, with the cost value to the destination found in the source's cost table. Any neighbor hearing the broadcast message and having a routing cost for the destination less than that sent by the source (and therefore presumably closer to the destination), waits for a random period and then rebroadcasts the message, listing its own routing cost. Acknowledgment is passive; the acknowledgment to the transmitting device is detection of the retransmission by a neighboring device. All others ignore the message, except to record the routing cost and destination sent by their neighbor, which they use to update their cost tables. In this way, the message slides down a *cost gradient* to its destination. The GRAd algorithm requires each network device that may be a message destination to flood the network upon network association with a special announcement message, so that it will appear in the cost table of other network devices. Alternatively, specialized route request messages can be broadcast by devices with messages for destination nodes for which they have no cost value stored.

- *Ant algorithms*: Because they also are ad hoc and self-organizing, biological analogies to the routing problem have also been explored. Of particular interest is the communication paradigm of ants. Ants communicate by placing a pheromone trail along the ground. If, for example, a wandering ant finds food, it returns to the nest, placing the food pheromone on the ground as it does so. Other ants detect this trail, arrive at the food, and return to the nest, laying down pheromone trails of their own that diffuse together into a wide trail. The pheromone evaporates quickly; an unused trail becomes undetectable to other ants in a short time. In this way, the ants quickly develop a single path to the food that is used by almost all ants.

Because this use of pheromones is a distributed communication system in which simple behavior of individual ants results in complex behavior of the larger network, it has clear appeal to the network routing problem, especially in the case where there is a single message sink, but possibly many message sources. One approach is to model the information sink (i.e., the PAN coordinator, the destination of all messages) as the nest, the network devices sourcing messages as the food, the collection of intermediate (relaying) network devices as the ground between food and nest, and the messages themselves as the ants. The PAN coordinator periodically transmits a special announcement message (containing a number-of-hops-taken field), which is passed from network device to network device by unicast transmissions. At each network device, the device stores in a table the source identity, the identity of the neighbor from which it received the message, and a time stamp. The network device then increments the number-of-hops-taken field in the message and forwards it to a different neighbor. Network devices perform the pheromone evaporation function by removing old entries from their tables.

When a real message is generated in the network, destined for the PAN coordinator, if the source has not received the special announcement message from the PAN coordinator recently (so that an entry is not present for it in its table), it sends the message to a random neighbor. This process is repeated until the message is sent to a network device that has a table entry for the destination. That device then sends the message to the neighbor from which it received the special announcement message, as recorded in its table. The message is now "on the pheromone trail," and it will eventually arrive at its destination.

The ant communication paradigm has appeal for particular networks in which it can be guaranteed that there will be only a few network devices acting as message destinations, because each destination must transmit regular announcement messages. Examples of such networks include wireless sensor networks, which typically have many source devices (sensors) and few destination devices.

The Cluster Network Topology

The previous algorithms notwithstanding, routing in a flat network has unavoidable limitations for many applications, and it would be preferable to have a network in which the routing algorithm was assisted in some way by the logical structure of the network, especially as the number of devices in the network becomes large. (Note that the physical structure of the network may still be ad hoc, however). This structure could eliminate, for example, much of the announcement message passing of the ant algorithm, and the exchange of very large routing tables inherent in the Bellman-Ford algorithm.

One approach to this problem suitable for many applications is the cluster network. In a cluster network, there is the concept of a parent-child relationship between network devices (Figure 6–5). The network forms, as do all IEEE Std 802.15.4 networks, with the PAN coordinator as the first device in the network. When a new device associates to the PAN coordinator (and therefore to the network), it becomes the child of the PAN coordinator, and the PAN coordinator becomes the parent of the new device. Should a second device come into range of the first network device (but, perhaps, out of range of the PAN coordinator), the second device may join the network as a child of the first device. Network devices may have many children and grandchildren, but only one parent (Figure 6–6).

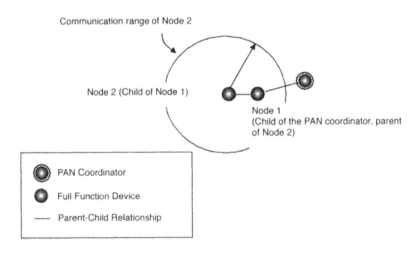

Figure 6–5: Formation of a cluster network

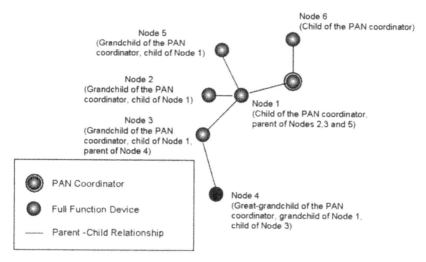

Figure 6–6:   Parents, children, and grandchildren in a cluster network

Discovery of the network by new devices is made possible by the transmission of beacons by each network device (including the PAN coordinator). When a new device appears, it might hear several beacons and have several potential parents from which to choose. If (as is often the case) it is preferable to connect network devices as close to the PAN coordinator as possible, information may be placed in the beacon payload to assist the prospective network member in the parental selection process.

The structure of the cluster network is controlled to some extent by the PAN coordinator, which retains authority over network association (regardless of which network device a prospective member may contact). The PAN coordinator may prohibit, for example, a prospective member from joining the network at a device distant (many hops away) from the PAN coordinator, while allowing other prospective members to join closer to the PAN coordinator, to encourage a flatter cluster structure and control the latency of messages sent in the network.

One advantage of cluster networks is that it is straightforward to perform a periodic network status update, so that the PAN coordinator can be made aware of any broken communication links or missing network devices anywhere in the network. One way to accomplish this update is to have the PAN coordinator generate a status-update-request message and send it to all network devices that are not parents (i.e., all devices at the ends of the cluster branches). When received, every parent device relays this message to all of its children, so that all network devices either receive or relay this message. When a network device without children receives the status update request message, it replies with a status-update-response message, a very short message containing its network address sent to the PAN coordinator. Intermediate devices on the return path to the PAN coordinator eavesdrop on this message as they relay it, gaining information about network devices in their cluster, but perhaps out of range. They then append their information to the message (perhaps consolidating the messages received from several children) and forward it to their parent. This procedure is shown in Figure 6–7.

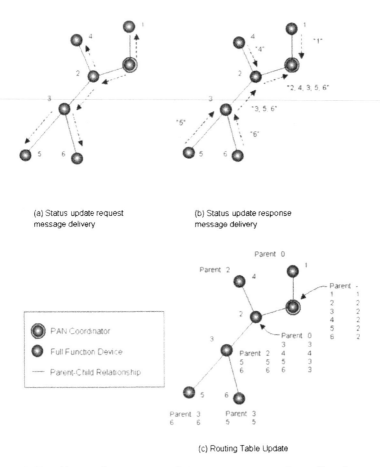

(a) Status update request message delivery

(b) Status update response message delivery

(c) Routing Table Update

Figure 6–7:   Network status update messages and routing in a network

A variant of this procedure is to require the status update response message to include the complete routing table of each network device. While lengthening the response message, this message can make the PAN coordinator (and, in fact, any eavesdropping network device) aware of alternative routes to the same destination device, which could improve network reliability if a network device or communication link fails.

Routing

Because their network devices have an understanding of network connections in their local area, routing in cluster networks can be more efficient than in flat networks. One efficient way to route in cluster networks is to use information gained in eavesdropping the status update response messages. By doing so, eavesdropping devices have a list of network devices downstream of the device from which the status response message was received. This information can be stored in a routing table.

A general routing algorithm for cluster networks is shown in Figure 6–8. The algorithm assumes that the network device stores a routing table, which consists of entries from two sources:

- Entries added by eavesdropping on devices within reception range. One of these entries will be for the parent of the device. These entries make up the neighbor list, the list of all devices to which messages may be passed without intermediate relay.

- Entries added by eavesdropping on network status reply messages. These messages inform the device of all devices in the network downstream that is away from the PAN coordinator.

Following the algorithm shown in Figure 6–8, a network device checks first if the destination device has an entry in its routing table. If an entry is found, the message is passed to the appropriate device. If an entry is not found, the message is sent to the parent, which, because it is closer to the PAN coordinator may have additional routing information by virtue of eavesdropping on network status response messages from a larger number of devices. In the extreme, a message may be relayed all the way to the PAN coordinator, but delivery is assured nevertheless.

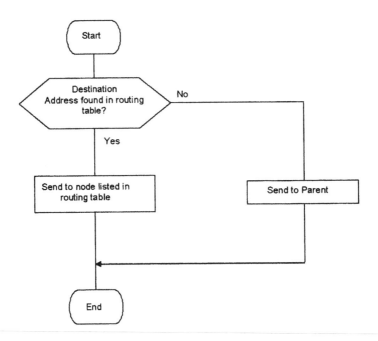

Figure 6–8: A message routing algorithm for cluster networks

A significant advantage of cluster networks for wireless sensor network applications is the small size of the network device routing tables, compared to those of flat networks. In general, for flat networks, a table entry is required for every potential destination; for a cluster network, the routing table may be much smaller because destinations not found in the table are still routed, through the parent. A small table size reduces the memory requirement of the network devices, which can be a major factor in their product cost.

A disadvantage of cluster networks is the nonuniform distribution of message traffic among the network devices. Some devices, especially those logically close to the PAN coordinator, might have significantly higher traffic than devices further away from the PAN coordinator, leading to unequal battery life among network devices. If the power supply of a device becomes exhausted, a network partition could occur. Several methods have been proposed to mitigate this effect, including the rotation of PAN coordinator duties among alternative network devices and the use of multiple routing table entries for the same destination.

The Cluster-Tree Network Topology

The address space of IEEE Std 802.15.4 MAC can support a large number of network devices. However, as the number of network devices continues to grow, even the routing tables of devices in a cluster network may grow to an impractical size, because a device's routing table must contain an entry for each of its offspring (child, grandchild, etc.).

To address this issue, hierarchy may be employed. A large network may be broken up into several, smaller clusters, connected in a hierarchical tree. An example of such a cluster-tree network is shown in Figure 6–9.

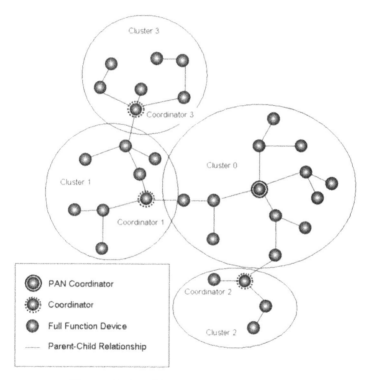

Figure 6–9:   The cluster-tree network

The large network shown in Figure 6–9 is composed of four smaller clusters, each with a cluster head, or coordinator. The coordinator for Cluster 0 is, of course, the PAN coordinator. If the PAN coordinator has a set of rules governing network formation, the clusters may be formed as a natural part of the network formation process. These rules may take the form of a limit on the number of devices in a cluster, a limit on the number of hops any device may be from a coordinator, or a more sophisticated algorithm. Should the attempted association of a candidate device to a network device violate these rules, the PAN coordinator may allow the device to join, but only as the coordinator of a new cluster.

The logical, short address of a network device in a cluster-tree network is now a hierarchical address consisting of two parts, a cluster identifier and a network device identifier. The network device identifier of a coordinator is always zero; the cluster identifier of the PAN coordinator is always zero.

In the cluster-tree network, the PAN coordinator may still transmit network status request messages. One way to do so is to have the network status request messages route, as in the cluster network, to the ends of the branches (the "leaves" of the tree). The network status response message is also handled in the same way, except that the device identifiers in the message are deleted by the eavesdropping coordinators; i.e., network status response messages leaving the coordinators contain only the identity of the cluster. This process takes advantage of the address hierarchy to reduce the size of the network status response message, which, in a large network, may otherwise be impractical. Note that, as in the cluster network, information extracted from the routing table of each network device may be included in the network status response messages to exploit routing redundancies.

Routing

A possible routing algorithm for a cluster-tree network is shown in Figure 6–10. In this algorithm, a device first checks to see if the cluster identifier of the destination is in its routing table. If not, the destination is an unknown cluster and the device routes the message to its parent.

However, if there is a cluster identifier match in the routing table, the device then checks those table entries with a cluster identifier match for a network identifier match. If one is found, the message is routed according to the table entry. If one is not found, the destination is an unknown network device in a known cluster, and the device must choose which network device to use to relay the message to the cluster by some set of arbitration rules.

If the cluster identifier of the destination is the same as that of the network device attempting to route the message, as in the cluster network, the message should be routed to the parent of the network device. However, the arbitration rules governing the relay of messages to other clusters can be significantly more flexible, and may consider, for example, the time a routing table entry was made—routing the message via the network device most recently added to the table, in the belief that the route is the most up-to-date and most accurately reflects the present state of the network. Other considerations are possible, many of which are optimal for particular applications.

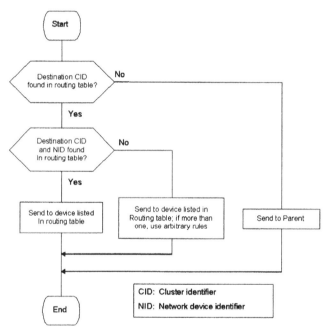

Figure 6–10:   A cluster-tree routing algorithm

THE NETWORK TOPOLOGY DECISION

Because IEEE Std 802.15.4 is capable of supporting a large number of ad hoc network topologies, an application designer employing IEEE Std 802.15.4 must select an appropriate network topology for the intended application. Further, if the selected network topology is a peer-to-peer topology in which multihop communication is planned, the routing algorithm must also be selected.

As an aid to this decision-making process, Table 6–1 summarizes the major network types described in this chapter, including their strengths, weaknesses, and potential applications. Note that Table 6–1 is only a guide, intending to establish general trends; almost any of these networks may be used success-fully in almost any application, albeit with variations in network performance parameters such as message throughput and latency and network device per-formance parameters such as power consumption.

> The network types listed in this section are only examples; many other topolo-gies and routing algorithms are possible.

Table 6–1: Network topology comparison

Network Type		Strengths	Weakness	Possible Applications
Star		Low message latency, centralized network control	Can cover only a limited physical area (single-hop communication)	Home automation, PC peripherals
Peer-to-Peer		Can cover a large physical area (multi-hop communication)	Higher message latency	Wireless sensor networks, industrial control and monitoring
	Flat	Simple network devices	Does not scale well as the number of potential destination devices increases	Wireless sensor networks
	Cluster	Supports a larger number of potential destination devices	Uneven power consumption among network devices	HVAC systems
	Cluster Tree	Can support very large networks	Network maintenance overhead	Industrial control and monitoring

Part III

Chapter 7   System Design Considerations

...a systemic view

IEEE Std 802.15.4 was created to address applications that could not be satisfied with existing WLAN and WPAN communication protocols, primarily due to the cost, size, and/or current drain of their implementations. This chapter describes system design considerations needed to take advantage of the capabilities of IEEE Std 802.15.4 and achieve low cost, size, and current drain implementations of the standard.

To reduce cost, size, and current drain of its implementations, IEEE Std 802.15.4 was designed for a high level of integration, with a minimum number of external parts. Several features in the protocol help achieve this goal.

DIRECT SEQUENCE SPREAD SPECTRUM

The use of direct sequence spread spectrum (DSSS) enables an IEEE Std 802.15.4 implementation to be largely a digital, with only a relatively small number of analog circuits. This design "future-proofs" the standard; the evolution of the standard as improvements in integrated circuit lithography will increase digital circuit density, leading to even lower implementation costs. DSSS offers other implementation advantages. The DSSS processing gain provides rejection to interfering signals, reducing the requirements on the channel filter. The ability to easily design orthogonal, multilevel signals with DSSS enables the IEEE Std 802.15.4 2.4 GHz physical layer to have simultaneously a relatively fast data rate of 250 kb/s (to finish transmission and return to sleep quickly) and a relatively low symbol rate of 62.5 kSymbols/s (to minimize current drain while active); both of these features improve average power consumption. Further, in some regulatory jurisdictions, some type of spreading is required, and DSSS enables faster network discovery and synchronization than does Frequency Hopping Spread Spectrum (FHSS). For these reasons, DSSS was chosen over FHSS and narrowband techniques.

HIGH CHANNEL SEPARATION/MODULATION BANDWIDTH RATIO

IEEE Std 802.15.4 has a high ratio of channel separation to modulation bandwidth. That is to say, adjacent channels are far away from the reference channel, compared to its width (see Figure 7–1). The first null in the transmitted IEEE Std 802.15.4 spectrum in the 915 MHz band is 600 kHz away from the signal center, with 1.4 MHz separation to the edge of the adjacent channel signal; in the 2.4 GHz band the first null is 1.5 MHz away from the signal center, with 3.5 MHz separation. In both cases, the separation-to-bandwidth ratio is 1.4/0.6 = 3.5/1.5 = 2.33. However, IEEE Std 802.15.4 requires very little selectivity to the adjacent channel, only the alternate channel, resulting in an effective ratio of 3.4/0.6 = 8.5/1.5 = 5.67. In comparison, the transmitted IEEE Std 802.11b/g spectrum, in which the first null is 11 MHz away, with 14 MHz to the edge of the next nonoverlapping adjacent channel, has a ratio of only 14/11 = 1.27; IEEE Std 802.11b/g requires 35 dB of selectivity to this channel. IEEE Std 802.15.4 channel filter stopbands are therefore well below the appropriate passband. A high ratio of channel separation to modulation bandwidth has the following beneficial effects:

- Receiver types that employ integrated selectivity, such as low-IF and zero-IF architectures, are possible. Discrete surface acoustic wave (SAW), ceramic, or crystal filters are not needed, reducing implementation size and cost.

- For low-IF and zero-IF receiver implementations, the low-pass cut-off frequency of the channel filter may be raised, reducing the size and cost of the integrated filter.

- Fewer poles of channel filtering may be used, also reducing the size and cost of the filter.

- With fewer poles of filtering, and a higher corner frequency, the time required to settle turn-on transients in the channel filter may be reduced, enabling a faster receiver warm-up, minimizing battery power lost during the warm-up period.

- The relatively large channel separation means that a high synthesizer reference frequency may be used. Because synthesizer lock time is related to the frequency at which the synthesizer's phase detector operates, a higher reference frequency enables shorter synthesizer lock times, which allows shorter warm-up times and therefore improves battery life.

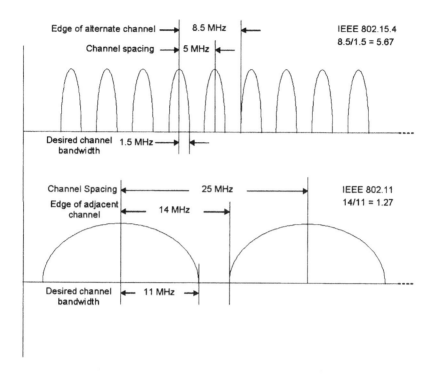

**Figure 7–1:   Comparison of IEEE Std 802.11 and IEEE Std 802.15.4
2.4 GHz PHY selectivity requirements**

RELAXED TRANSMITTED ERROR VECTOR MAGNITUDE (EVM) REQUIREMENTS

With the relaxed reception selectivity and transmitted error vector magnitude (EVM) requirements of IEEE Std 802.15.4, the sideband noise specifications of the receiver local oscillator and transmitter oscillator can be relaxed, perhaps to the point that a low-cost, fully integrated oscillator may be employed. This integrated oscillator, which could be of the inductor-capacitor (LC) or (preferably) the resistor-capacitor (RC) type, would eliminate the large and costly external voltage controlled oscillator (VCO) resonator, and other associated passive components, that would otherwise be needed.

CONSTANT-ENVELOPE MODULATION

A major factor leading to reduced transmit power consumption is the use of constant-envelope modulation in the form of half-sine-shaped O-QPSK in the high band. Transmitters must be designed, of course, to transmit the peak envelope power required by their modulation format; for simple circuits suitable for IEEE Std 802.15.4 implementations, this requires the amplifier dc bias power (which can be a significant fraction of the total transmitter power consumption) to be proportional to the peak envelope power. The average envelope power, on the other hand, is the significant parameter when determining range (given a fixed data rate). If a modulation format is chosen such that the peak envelope power is greater than the average envelope power, dc bias power must be spent in the transmitter commensurate with the higher peak power, even though the range achieved is that of the lower average power—clearly an undesirable situation for power-sensitive applications. (Such modulation formats, however, may have advantages, such as spectral efficiency, that make them suitable for other applications.) IEEE Std 802.15.4 therefore employs a modulation format having a constant envelope, in which the peak and average envelope powers are equal (a peak-to-average power ratio of one). An implementation of IEEE Std 802.15.4 may take advantage of this to employ a simple and relatively efficient transmitter.

RELAXED RECEIVER MAXIMUM INPUT LEVEL

The IEEE Std 802.15.4 receiver maximum-desired signal input level specification, –20 dBm, is significantly less than that of other WLAN and WPAN standards. An IEEE Std 802.15.4 system designer can therefore reduce the current drain of the receiver front end accordingly.

TIME AND FREQUENCY REFERENCE TRADE-OFFS

The design of the time and frequency reference subsystem of an IEEE Std 802.15.4 implementation can have a marked effect on overall system cost, size, and current drain. One possible approach to the design of a stand-alone IEEE Std 802.15.4 implementation employs two reference crystal oscillators, a high frequency oscillator for the RF frequency reference and a low-frequency oscillator for the protocol time base. In this design, the low-frequency (and therefore lower power) crystal oscillator for the time base runs constantly, providing protocol timing information. The high-frequency (and therefore higher-power) crystal oscillator for the RF frequency reference is then started immediately before a transmit or receive period and is turned off immediately afterward, thereby minimizing its power consumption. If the required warm-up time of the RF frequency reference is low enough so that it is not a significant factor in total energy consumption for the application under consideration, this approach can produce low average power consumption, at the cost of requiring two external crystals.

A second approach employs only a single, high-frequency crystal oscillator. In this scenario, the oscillator is used for both the RF frequency reference and the protocol time base, and it is therefore never turned off. (The oscillator output is divided down to a low frequency to be used as the protocol time base.) When needed as the RF frequency reference, its output is used directly; because it is already operating, no warm-up period is needed and therefore no power is lost there. This second approach is lower cost than the first, because only one crystal is needed; however, because a higher-frequency oscillator, plus frequency division circuitry, is used for the time base, the average power consumption may be higher. The system designer must evaluate this trade-off based on the expected applications of IEEE Std 802.15.4 implementation.

SINGLE-CHIP VERSUS MULTIPLE-CHIP IMPLEMENTATION

An integrated circuit (component) manufacturer entering the IEEE Std 802.15.4 market is faced with the decision of whether its design should be a two-chip design, in which IEEE Std 802.15.4 RF transceiver is on a separate chip from the protocol handler (Figure 7–2a), or a one-chip design, in which the RF transceiver and protocol handler are integrated on the same die (Figure 7–2b). A one-chip design, because it has the highest level of integration and fewest number of IC packages, might have the smallest implementation and lowest component cost, and it could be optimum for stand-alone applications that do not have an available host processor (as described in the following sections). However, a two-chip design may employ more optimal IC processes that are better suited to the circuit requirements (e.g., BiC-MOS for the RF transceiver and CMOS for the protocol handler), and it could result in a reduced time to market compared to a chip in a compromise integrated circuit (IC) process chosen for the one-chip design. In addition, if a wide product line of stand-alone IEEE Std 802.15.4-enabled products is envisioned, a two-chip design might allow the substitution of multiple processor types as the protocol handler IC, having different amounts of memory and capable of economically supporting a range of applications of varying complexity. The two-chip design also allows flexibility in the choice of RF transceiver with a single protocol handler. For example, one might want to enable either a dedicated IEEE Std 802.15.4 RF transceiver or one capable of servicing multiple protocols, for example both IEEE Std 802.15.4 and IEEE Std 802.11a. A compromise alternative to this decision is to employ two IC die, but placed in a single IC package.

OEM IMPLEMENTATIONS

The system design decisions faced by an original equipment manufacturer (OEM) are equally interesting. One scenario involves the addition of IEEE Std 802.15.4 capability to a larger system; for example, when a general-purpose host processor is available to run an IEEE Std 802.15.4 protocol stack, and a dedicated IEEE Std 802.15.4 protocol-handling processor is not used. In this case, shown in Figure 7–2c, the cost of the dedicated processor may be saved by employing a smaller chip performing only IEEE Std 802.15.4 RF transceiver functions—half of the two-chip design just described.

However, software integration is made more difficult, especially if the host processor is running other real-time applications, because the performance of many protocol functions cannot be delayed. Interrupt handling and context switching, for example, must be carefully evaluated to ensure that enough processing capability (e.g., million instructions per second [MIPS] random access memory [RAM]) is always available. Also, the throughput and latency performance of the communication link between the IEEE Std 802.15.4 transceiver and host processor must be understood. Often the worst-case scenario, when the host processor is most heavily loaded, is difficult to predict and test, especially for embedded systems. If the software integration problem becomes untenable, a one-chip IEEE Std 802.15.4 implementation may be used to off-load the host processor, as shown in Figure 7–2d.

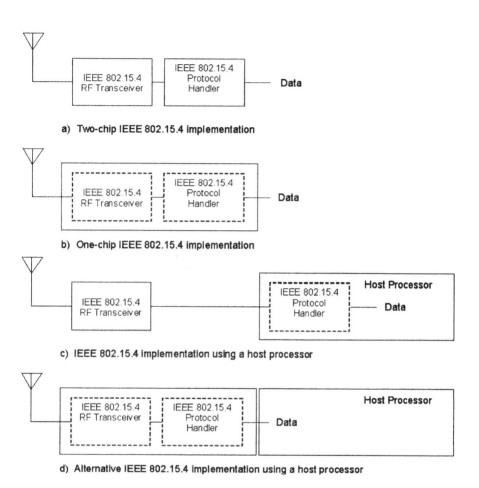

a) **Two-chip IEEE 802.15.4 implementation**

b) **One-chip IEEE 802.15.4 implementation**

c) **IEEE 802.15.4 implementation using a host processor**

d) **Alternative IEEE 802.15.4 implementation using a host processor**

Figure 7–2: Diverse IEEE Std 802.15.4 hardware implementations

Lightly loaded host processors, on the other hand, might have little difficulty meeting the real-time timing requirements of the protocol stack. However, unless specifically designed for the task, the host processor may not perform the bursty computation requirements of IEEE Std 802.15.4 protocol in a power-efficient manner; the average power consumption of the resulting IEEE Std 802.15.4 implementation may suffer as a result. The system designer could achieve the best compromise between implementation cost and power consumption by judiciously placing only those stack components on the host processor that best fits the needs of the intended application.

IEEE Std 802.15.4 was designed to be compatible with low-cost 8051-based or HC08-based microcomputers operating at a bus speed of only a few mega-hertz. This design involved a trade-off between the utility and flexibility of multiple protocol layers, and the software size reduction that occurs when the protocol is defined as a single layer. The compromise reached was to keep the layered architecture, to retain compatibility with other IEEE 802 standards, but simplify each protocol layer as much as possible.

> Even when the original design vision of IEEE Std 802.15.4 was to use small footprint microprocessors, today most implementations allow the radio device to be interface with 16-bit and 32-bit based architectures.

TIME AND POWER MANAGEMENT

IEEE Std 802.15.4 is designed to support a time base with a tolerance as great as ±40 ppm. This time base allows the use of very inexpensive reference crystals, minimizing cost. At 2.4 GHz, the standard also supports a beacon interval of $2^{14} \times 0.01536s = 251.6574s$, or over four minutes. This interval, combined with a minimum beacon PPDU length of 544 μs, enables a transmit duty cycle of $544 \times 10^6 / 251.56824 = 2.16 \times 10^{-6}$. Lower values are, of course, possible if beacons are not used at all.

However, a *receiving* device must be receiving at the time the beacon is sent, a time that is uncertain due to the imperfect accuracy of both transmit and receive time bases (Figure 7–3). Without time base regulation, the receiver must receive for a period of $2\varepsilon T_{beacon} + T_c$ s every T_{beacon} s, where ε is the sum of the transmit time base tolerance ε_{Tx} and receive time base tolerance ε_{Rx}; the lowest possible average duty cycle of a receiving device is therefore

$$2\varepsilon + \frac{T_c}{T_{beacon}} \qquad \text{EQ 7–1}$$

It is clear that, no matter how long the beacon period T_{beacon}, the duty cycle is limited by the attainable time base stability ε. (Neither transmit nor receive warmup times are included in this analysis.) For 40 ppm time bases at both transmitter and receiver, this effect becomes significant for $T_{beacon} > 0.5$ s or so. To achieve the very low duty cycles (and resulting long battery life) of which possible for devices employing IEEE Std 802.15.4 are capable, an implementation may therefore require a more stable time base than that specified in the standard. Although the simplest way to do this may be to replace, for example, a 40 ppm crystal oscillator with one of greater stability, this would usually increase the cost of the time base significantly—especially if stabilities below 5 ppm were required. A lower cost approach could be developed by recognizing that the time variation between sequential beacons from the same device is likely to be highly correlated. It is then possible for the receiving device to reduce the device duty cycle by tracking the relative received beacon time from beacon to beacon, and therefore employing a reception window smaller than $2\varepsilon T_{beacon} + T_c$.

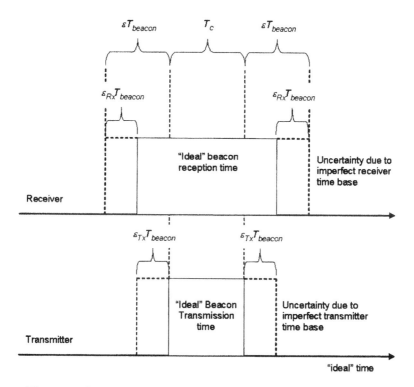

Figure 7–3:   Effect of imperfect transmitter and receiver time

Alternatively, a time base drift specification, limiting the allowable time difference between successive beacon periods, could have been specified in

IEEE Std 802.15.4, as has been done in other standards. It is clear that a drift specification is not needed for macBeaconOrder = 0; when T_{beacon} = 15.36 ms, any possible drift is insignificant when compared to the beacon PPDU length (544 μs at 2.4 GHz) plus warmup time (a total of at least, for instance, 600 μs). The drift time, even assuming an 80 ppm difference between devices, does not become significant when compared to 600 μs until one reaches macBeaconOrder = 7 (1.96608 s), when it is 157 μs.

It is also clear that a drift specification is not useful for macBeaconOrder = 14; a device could move from hot to cold during the four-minute beacon period, and the warm-up time estimate could only be improved by improving the accuracy of the time base, an undesirable option. Small, single-chip IEEE Std 802.15.4 implementations (perhaps with button cell batteries) might have extremely low thermal inertia; when moving between temperature extremes during a 16-second beacon period (macBeaconOrder = 10), for example, they could well vary by ±40 ppm. A drift specification could not be applied here, either, without adding a significant (i.e., costly) restriction on the hardware. In this regime, macBeaconOrder = 10 to 14, the receiving time must be specified based on the time base accuracy specification.

This leaves a small window in which a drift specification would be useful, macBeaconOrder = 7 to 9. It was decided that specifying drift only for three macBeaconOrder values was not appropriate for a simple protocol with low implementation cost, so a time base drift specification is not included in IEEE Std 802.15.4.

The previous paragraph shows the importance of time and power management in the design of IEEE Std 802.15.4 implementations, and the concern for them that went into the design of the standard. Note that, because the standard is designed to support duty cycles of 0.1% or less, the standby power consumption of the implementation can be the dominant factor in average power consumption (and therefore battery life). It is therefore rewarding to pay attention to standby power consumption during the design process. In addition, because active times are brief (as short as 544 μs for a minimum-length 136-bit beacon PPDU at 2.4 GHz), warm-up times can be a significant fraction of overall active time and can therefore affect the minimum obtainable duty cycle. This effect is important if maximum battery life is to be achieved.

If the application can support a star network configuration with a mains-supplied PAN coordinator, and especially if the PAN coordinator is to be the destination of all network traffic (for example, in a lighting network, in which several wireless switches control a lamp PAN coordinator), the other network devices may have an arbitrarily low duty cycle by operating the network without a beacon. In this mode, the PAN coordinator is receiving constantly; the network is in a constant contention access period.

The switches need not transmit or receive at all until some event (e.g., the toggling of a switch) causes a message transmission from a switch to the lamp. Because a lamp is typically connected to the mains, having a near-100% receive duty cycle is not an onerous requirement in this application, and it enables the designer to craft a system in which the switch battery life may be exceptionally long.

ANTENNAS

Because many IEEE Std 802.15.4 products are expected to be physically small and sensitive to product cost and power consumption, antennas can be crucial to a successful design. The physically small antenna required in many applications may be inefficient without careful attention to the design of both the antenna and its placement in the product. An inefficient antenna can significantly reduce both the transmission and reception range, which can be compensated for only by increasing transmit power and receiver sensitivity—both options that can significantly reduce battery life.

There is a trade-off to be made among the physical size, instantaneous fractional bandwidth BW, and the maximum achievable radiating efficiency η of electrically small antennas (i.e., those with maximum physical dimensions significantly less than their operating wavelength λ). Specifically, efficiency is represented by the following equation [24]:

$$\eta = \frac{2\left(\frac{2\pi r}{\lambda}\right)^3}{BW} \qquad \textbf{EQ 7–2}$$

where $2r$ is the maximum dimension of the electrically small antenna (i.e., the diameter of the smallest sphere that can completely contain the antenna).

From this equation, two important relationships may be noted:

- The antenna efficiency is proportional to the ratio of the antenna size and the operating wavelength, *cubed*. Therefore, all else equal—including the physical size of the antenna—operation at a higher frequency (a lower wavelength λ) improves antenna efficiency, as does increasing the size of the antenna while keeping the frequency of operation constant. If, however, the antenna size scales with wavelength (as is the case for a half-wave dipole, for example), its efficiency is unaffected by the frequency of operation.

- Attainable fractional bandwidth BW is inversely proportional to attainable efficiency η. In most applications, the efficiency of even well-designed small antennas for any of IEEE Std 802.15.4 bands is low enough that the bandwidth required is easily achieved.

In addition to the antenna size $2r$, strong circulating (i.e., non-radiating) fields exist near an antenna. These fields vary as d^{-2} and d^{-3}, where d is the distance from the antenna. The fields are therefore very strong close to the antenna, and significant out to a radius of at least λ/10. Any material in this region, including batteries, circuits, displays, and so on, produces additional loss that can significantly affect antenna efficiency. A designer needs to plan for a keep-out region in the product design of 1.2 cm or so for 2.4 GHz operation, and approximately 3.5 cm for 868 MHz operation. The more material in this region, the lower the antenna efficiency will be. When keep out regions of this size are not possible, the increased loss should be considered in the implementation design. Note that the effect of nearby materials can counteract the increased efficiency of a larger antenna over a smaller one, if the larger antenna must be placed near a lossy object while the smaller antenna is kept away from it.

This theory must be matched with the practical realities of antenna design for small, low-cost products. Because any volume taken by an antenna inside a product is not available for other uses (the placement of circuits, batteries, etc.), it is nearly always better to minimize it. Several factors affect this product design decision:

- *The frequency of operation*: As stated previously, for constant efficiency the antenna size scales with the operation wavelength; operation in the 2.4 GHz band would therefore consume less antenna volume.

- *Products for multiple bands*: If a single product design is intended to be used for both 868 MHz and 2.4 GHz operation, for example, space must be made available for the larger 868 MHz antenna.

- *Internal noise sources*: Because of their proximity to the antenna in a small product, even relatively weak noise sources can play havoc with a wireless transceiver. Typical sources are digital circuits (especially the busses of microprocessors and microcomputers, and serial data interfaces) and switching power converters. These components, and the circuit board traces associated with them, should be placed as far as possible from the antenna, something that becomes more difficult as the antenna size increases. When possible, of course, it is better to arrange operation of the digital circuits so that they are not clocked when the transceiver is active.

- *Use of external antennas*: In general, external antennas (those outside the product housing) perform better electrically; they are typically away from lossy materials and have fewer size restrictions; so they may be larger. If a directional antenna is needed (perhaps for a long-range application), an external antenna is often the best choice; they may also be required in applications such as wireless home appliances (e.g., a refrigerator) in which the radio transceiver is enclosed in a metal housing. External antennas also maximize the internal volume available for other product components. However, they may cost more than internal antennas (especially if RF connectors are needed) and, because they lack the protection of the product housing, they may be prone to mechanical failure due to environmental factors, handling, and so on. The loss of the feedline between the antenna and transceiver must also be considered in the product design.

If an internal antenna is the decision, there are several design choices available. The first choice is often to employ a circuit board trace as an antenna, certainly the cheapest approach, because no antenna components need be purchased, handled, or placed on the board. However, it is not free, because circuit board area must still be purchased for it. On the negative side, circuit board antennas tend to have significant loss, because of their resistance (caused by the thinness of the copper trace) and their proximity to lossy materials such as the circuit board material itself, nearby ground planes, traces, and other circuit components.

Wire antennas, such as loops and dipoles, allow the antenna to be placed above the plane of the circuit board, improving its efficiency, and they are often a good compromise between the performance of external antennas and the cost of circuit board trace antennas. Wire antennas might require tuning for best performance because of variations in mechanical shape and size that affect their frequency of resonance. They also might need nonmetallic physical supports in certain applications.

Specialized antenna components, composed of ceramic materials, are also available. They offer a smaller physical size than that attainable with wire antennas and do not require tuning; however, they are more expensive than wire antennas.

PRODUCT DESIGN FLEXIBILITY

IEEE Std 802.15.4 is designed to have the flexibility to serve in a wide variety of applications—from personal computer peripherals and toys to home automation and industrial controls. These applications, however, could have very different cost, size, and current drain goals, and the ideal design for one may be unsuitable for another. Figure 7–4 is a spider chart showing the relative importance of size, cost, and current drain for three possible IEEE Std 802.15.4 applications. Note that these determinations are examples only and will vary greatly, depending on implementation details and the specific applications chosen within these broad categories.

In this illustrative example, cost is of primary importance in the home automation application, whereas current drain is of the least importance—possibly because the network has access to mains power. Similarly, current drain is of primary importance in the industrial control application, whereas size is of the least importance—possibly because the application is a large industrial motor controller. Finally, size is the primary consideration in the PC peripheral application—for example, a wireless mouse or joystick—whereas cost is of the least importance. The application designer must make the right trade-offs in these three dimensions to optimize an IEEE Std 802.15.4 product.

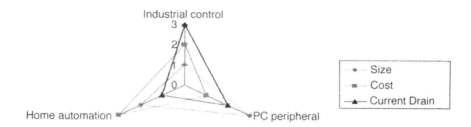

Figure 7–4: Example of relative importance of performance parameters for different applications

Chapter 8 Real-World Issues

...a touch of reality

Users, implementers, and system integrators of IEEE Std 802.15.4 should be aware of some factors that could affect the performance of their systems resulting from interaction with the "real world." Two of these factors are the coexistence of IEEE Std 802.15.4 products with other users of the radio frequency spectrum and the interfacing of different communication protocols.

COEXISTENCE

Because the bands used by the two IEEE Std 802.15.4 PHY layers are unlicensed in nearly all countries, devices employing other communication protocols share the bands. In the 868 MHz band, for example, there are proprietary low-data-rate protocols; in the 915 MHz band, proprietary protocols and cordless telephones; and in the 2.4 GHz band, cordless telephones, microwave ovens, WLANs, and other WPANs. In most regulatory regions, devices operating in these unlicensed bands must accept interference to them caused by other services operating in the same band, and their users do not have legal redress for any performance degradation caused by the other services. Although this might be the regulatory reality, it is not the market reality; users purchasing new devices reasonably expect that not only will the new devices operate as they want, but also that the devices also will not disrupt the operation of existing products. A device not meeting these two criteria could have a difficult time remaining on the market. Coexistence is therefore an important economic issue.

Several features of IEEE Std 802.15.4 were designed with coexistence with other services in mind. Although the issue can be considered as two problems (the protection of other services from interference caused by IEEE Std 802.15.4 transmissions, and the protection of IEEE Std 802.15.4 implementations from interference caused by the transmissions of other services), in practice most solutions to one of these problems are also solutions to the other. Some features of IEEE Std 802.15.4 that reduce potential interference to other services are listed in the following paragraphs. Much of this list is derived from the Coexistence Annex of the standard.

- *Low Transmitted Power*: Although radio regulations in many parts of the world (including Part 15.247 of the FCC rules [25] allow operation in the 915 MHz and 2.4 GHz bands with relatively high output power (as much as 1 W in some cases), the typical IEEE Std 802.15.4 device will likely operate with much lower transmit power. A key metric of IEEE Std 802.15.4 is cost, and achieving greater than 10 dBm transmit power in a low-cost system-on-chip (SOC), although feasible, will be economically disadvantageous. Furthermore, FCC and European regulations (ETSI EN 300 220-1 V1.3.1 [6], CEPT Recommendation 70-03 [5], and ETSI SN 300-328 [7] for out-of-band emissions make it difficult to employ powers above 10 dBm without additional, expensive filtering of the transmitted spectrum. These factors will limit the distribution of devices with greater than 10 dBm transmit power to a few specialized applications.

- *Channel Alignment*: A concern in many applications of IEEE Std 802.15.4 that employ the 2.4 GHz PHY is coexistence with WLANs using the popular IEEE Std 802.11, particularly those WLAN installations in which the maximum of three non-overlapping WLAN channels are employed. Figure 8–1 shows the alignment between the North American non-overlapping IEEE Std 802.11b/g channels and IEEE Std 802.15.4 channels. There are four IEEE Std 802.15.4 channels that fall in the guard bands between (or above) the three IEEE Std 802.11b/g channels (n =15, 20, 25, 26 for North America, n = 15, 16, 21, 22 for the European non-overlapping WLAN channels). Although the energy in this guard space will not be zero, it will be lower than the energy within the channels, and operating an IEEE Std 802.15.4 network on one of these channels will minimize interference between systems. When performing dynamic channel selection (controlled by the upper layers), either at network initialization or in response to channel impairment, an IEEE Std 802.15.4 device will scan a set of channels specified by the ChannelList parameter in the MLME-SCAN.request primitive. For IEEE Std 802.15.4 networks that are installed in areas known to have high IEEE Std 802.11b/g activity, the

ChannelList parameter can be defined as one of the previous sets (either [16, 23, 28, 29] or [16, 18, 24, 25], as appropriate) to maximize the coexistence of the networks.

- *Channel Selection*: Before the network forms, the candidate PAN coordinator scans all channels in its channel list, to identify IEEE Std 802.15.4 PANs already in existence. If a suitable PAN is identified, at the discretion of the network layer the candidate may choose to join that PAN, rather than create a second PAN. This behavior minimizes the number of PANs existing in a band, thereby reducing potential interference to other services. If no acceptable PANs are identified, however, the candidate PAN coordinator may choose to start a new PAN. It selects an unoccupied channel for the new PAN from the list of channels in the ChannelList parameter (which can be determined by an energy detection channel scan), a second behavior that avoids transmission on frequencies occupied by other services.

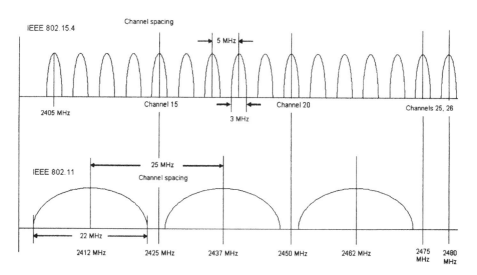

Figure 8–1:   Alignment of the North American IEEE Std 802.11 nonoverlapping channel plan and IEEE Std 802.15.4 2.4 GHz PHY channels

If interference appears on the channel occupied by the PAN, the upper layers of the PAN coordinator execute a dynamic channel selection algorithm (not defined in IEEE Std 802.15.4). The PAN coordinator scans the available channels in its channel list and selects a new channel for the PAN. It then returns to the original, impaired channel and makes a broadcast message to the PAN, identifying the new channel to which the PAN is to move. This procedure helps IEEE Std 802.15.4 systems avoid causing interference to, and receiving interference from, other services.

- *Clear Channel Assessment*: The CSMA-CA channel access mechanism in IEEE Std 802.15.4 performs a Clear Channel Assessment (CCA) before transmission. The PHY requires that at least one of three CCA methods be used—energy detection (ED) over a certain threshold, detection of a signal with IEEE Std 802.15.4 characteristics, or a combination of the two. Use of the energy detection option improves the coexistence behavior of IEEE Std 802.15.4 with other services by allowing transmission backoff if the channel is occupied by any device, regardless of the communication protocol it uses.

- *Use of Spread Spectrum Techniques*: The direct sequence spreading technique used by IEEE Std 802.15.4 provides some protection for users of narrowband (i.e., less than 200 kHz wide) communication protocols, by spreading the 250 kb/s data transmission over a bandwidth greater than 2 MHz, thus reducing the energy of IEEE Std 802.15.4 transmission present in any narrowband channel, protecting these services. Interestingly, even though its signal is spread, because of its low data rate, IEEE Std 802.15.4 transmission bandwidth is comparable to that of IEEE Std 802.15.1 (Bluetooth); as a result an IEEE Std 802.15.4 PAN interferes with a Bluetooth PAN much as a second Bluetooth PAN would—affecting only three of Bluetooth's 79 hops.

- *Link Quality Indication*: IEEE Std 802.15.4 PHY layer standard specifies a Link Quality Indication (LQI). The LQI is performed on each received packet, and it may be implemented using received signal strength, a signal-to-noise ratio estimation, or a combination of these. LQI can be used to detect channel impairment caused by interference on a packet-by-packet basis, providing real-time information to the upper layers of the channel condition, enabling the device to make an informed decision in dynamic channel selection.

The WirelessHART standard makes use of a clever combination of the coexistence measures provided by IEEE Std 802.15.4 to offer a high level of reliability in face of the radio-frequency challenges of industrial environments. Part IV of the book elaborates on the details how industrial-grade coexistence is achieved.

IMPLEMENTATION OF COLOCATED TRANSCEIVERS

An implementer of a system employing two RF transceivers, such as a cell phone incorporating a WPAN, should be aware of interactions that might occur between the two devices. Many applications require such a design, including those in which the cell phone incorporates a PAN coordinator and is used as the gateway device between the WPAN and the Internet. These interactions can be due to both radiated and conducted effects.

A primary concern when placing two RF transceivers in the same product (or, indeed, in close proximity) is the effect the radiated transmitted noise of one device may have on the receiver of the second device. All transmitters transmit, in addition to the actual signal, some amount of wideband noise, often for many hundreds of megahertz around the signal. Ordinarily, because the amplitude of this noise is so low, it is of no consequence, and it is not economically justifiable to insert filters or other circuits to reduce it further. However, if a receiver (or, more specifically, a receiving antenna) is brought very close to the transmitting antenna, the noise can become significant. In this case, the noise may affect the apparent receiving sensitivity of the second device, even though the frequencies of operation of the two transceivers differ greatly. If the communication protocols of the two transceivers are independent, so that the transmission of one device cannot be prohibited while the second device is receiving, special measures, in the form of adding selectivity to the transmitter output, might have to be performed to enable successful operation of the two transceivers.

A second radiated effect is the effect of receiver blocking. Blocking, a strong signal effect (as opposed to a noise effect), occurs when a strong, unwanted signal enters a receiver and affects the dc bias of its circuits. This intrusion typically causes a loss of gain, leading to a loss of sensitivity. In a sense, this event affects the signal portion of the receiver detector's signal-to-noise ratio, whereas wideband transmitted noise affects the noise portion. Blocking can happen, for example, if a 600-mW cellular transmitter is co-located with a WPAN receiver that is biased for low-power operation. Blocking performance can be improved by adding selectivity before the receiver input, if the offending signal is outside the passband of the victim receiver or, usually less desirable, by increasing the dc bias of the receiver circuits (i.e., increasing power consumption).

A conducted effect that can arise when employing two transceivers results from the sharing of a power supply. In portable devices, this effect is often unavoidable, because the primary supply is a battery and there is rarely room in the size or weight budget for a second battery in the product. However, if one transceiver draws significant current from the shared supply (usually while transmitting), any voltage drop resulting from either internal or external resistance associated with that supply could affect the second transceiver. The textbook example of this phenomenon is the supply voltage modulation caused by a time division multiple access (TDMA) cellular transmitter, such as a global systems for mobile (GSM) transmitter, which, because it is at an audio rate, caused an audible tone in early GSM handsets. This problem can be corrected with suitable supply design, but the addition of a separate voltage regulator for the victim device is often required.

Part IV

Chapter 9 Introduction to WirelessHART

...leveraging 802.15.4 in process control applications

WirelessHART, or IEC 62591, is the only standard for industrial wireless network designed on IEEE Std 802.15.4 technology. The intended use of this technology is to enable monitoring and control applications for industrial process automation including pharmaceutical, oil and gas, pulp and paper, water/waste-water treatment, chemical, and other industrial plant applications. These applications demand high levels of reliability and security when compared to residential and commercial applications. In addition, industrial applications require operation over wide–spread areas with large amount of devices building the network.

For most networking systems, the loss of a few communication packets is normal. However, for industrial applications, such loss is unacceptable. Wireless sensing devices in industrial applications are expected to run without any interruption for several years, especially in continuous automated processes. As will be shown later, this significant reliability requirement is addressed by adding multiple system redundancies in the WirelessHART design.

Security requirements are also stringent in industrial networks, putting high emphasis in process data confidentiality and data integrity of the process data. For example, in simple tank level monitoring, it is critical to keep the level information confidential at the time of commercial negotiations among suppliers for competitive reasons. Similarly a level information of a vessel containing a toxic chemical shall be correct (not tampered). Otherwise, there is a risk that an overflow will harm workers and the environment. In summary, it is highly important for industrial process data to be transmitted with the highest level of security. Because a radio-based communication system cannot be physically secured as a wired system can be, it is mandatory to overlay information security technology protocols for wireless communication system in industrial applications.

Security in the minds of operators and designers of industrial control applications has the following meanings:

• To prevent the illegal or unwanted penetration of the network.
• Intentional or unintentional interference with the proper and intended operation.
• Inappropriate access to confidential information.

WirelessHART copes with the significant requirements for reliability and security without defining a complex system. Simplicity is a fundamental feature envisioned to augment the robustness of the design while allowing overall operation by unskilled labor.

WirelessHART also addresses the need of industrial plant operators to use standards-based technologies by leveraging IEEE Std 802.15.4. Standards enable the supply of similar devices from multiple vendors, enticing competition, lowering prices, and increasing quality.

In addition, different device types from different vendors having a common communication network capability will interwork and even interoperate together. A multiple-vendor approach guarantees a wide variety of device types. Finally, well maintained standards are stable over time and typically maintain backwards compatibility with previous versions, eliminating the issues associated with the use of proprietary technology.

An industrial plant needs typically thousands of devices as a basis for production automation. The coordination of this many devices poses a challenge because all measurements from each sensor shall be transmitted in real time, meaning that the data is transmitted and received within a predefined time. That constrains the communications to a specified latency. The latency for typical industrial process automation sensing devices ranges from 1 second to 100 seconds.

Characteristically, industrial process automation plants cover large areas, WirelessHART allows the communication over these large areas by providing meshing capability. Devices can forward data on behalf of other devices over large distances.

WirelessHART technology uses the 2.4 GHz ISM band provided by IEEE Std 802.15.4. Industrial applications have a global footprint, requiring a wireless spectrum availability worldwide. As introduced in the earlier chapters of this book, the use of this band brings the additional advantage that no frequency licensing is required. However, the technology shares the spectrum with other services based on proprietary technology or standards-based technologies such as ZigBee and 802.11/WiFi. The latter is of particular relevance, because IEEE Std 802.11 is widely deployed as a backbone network for many industrial applications. WirelessHART selected IEEE Std 802.15.4 because this standard was designed to provide a high degree of coexistence with IEEE Std 802.11.

> The WirelessHART standard was released on September 12, 2007. In March 2010, WirelessHART was adopted by the IEC as an international standard under the denomination IEC 62591 and by the European community as EN IEC 62591 on June 2010.

HIGHWAY ADDRESSABLE REMOTE TRANSDUCER (HART)

The HART—Highway Addressable Remote Transducer—Protocol is the global standard for sending and receiving digital information across analog wires between smart devices and control or monitoring industrial system. HART is the leading communication system for the industrial process automation area. The HART protocol was developed in the late 1980s and transferred to the HART Communication Foundation in the early 1990s.

HART is the industry *de facto* standard for process instrumentation, with over 26 million installed devices representing more than 50% of the global industrial sensing population, and a pedigree running back over 20 years. The HART standard has evolved over the years maintaining backwards compatibility, each update adds to the specification and takes nothing away; the current version (HART 7) introduced wireless technology for first time. WirelessHART is part of the HART 7 standard.

The HART Protocol implements layers 1, 2, 3, 4, and 7 of the Open System Interconnection (OSI) 7-layer protocol model. HART is a bidirectional communication protocol that provides data access between intelligent field instruments and host systems. A host can be any software application from technician's handheld device or laptop to a plant's process control, asset management, safety system, or other system using any control platform.

At the highest protocol layer, the HART communications protocol provides three classes of application commands:

- *Universal commands*: These commands apply to all sensing and control devices, including diagnostic and device tagging.

- *Common Practice commands*: These commands apply to a variety of sensors and actuators, mostly focused on calibration and dynamic range adjustment.

- *Device Specific commands*: These commands represent functions that are unique to each field device. The device manufacturer specifies these commands.

Table 9–1 shows some of the Universal and Common Practice commands defined in the HART protocol.

Table 9–1: Examples of HART Universal and Common Practice commands

Universal Commands
Read manufacturer and device type
Read primary variable (PV) and units
Read current output and percent of range
Read up to four predefined dynamic variables
Read or write eight-character tag, 16-character descriptor, date.
Read or write 32-character message
Read device range values, units and damping time constant
Read or write final assembly number
Write polling address

Table 9–1:   Examples of HART Universal and Common Practice commands (continued)

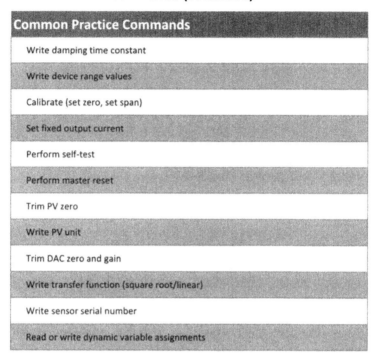

Common Practice Commands
Write damping time constant
Write device range values
Calibrate (set zero, set span)
Set fixed output current
Perform self-test
Perform master reset
Trim PV zero
Write PV unit
Trim DAC zero and gain
Write transfer function (square root/linear)
Write sensor serial number
Read or write dynamic variable assignments

WirelessHART Technology

Building on the rich tradition of HART, the HART Communication Founda-tion decided to develop a wireless technology that provides additional capa-bilities for reaching areas of plant that were not economical or are technically difficult to reach with conventional wired systems. At the same time, Wire-lessHART uses the traditional HART technology and products that are already in place by allowing these devices to be full participants in a wireless network. A WirelessHART device uses the same command structure from wired-based HART devices and works with the same software platform and associated tools.

Transmitting HART protocol messages over a generic wireless cloud is not feasible for industrial wireless applications. These applications require a level of timing control, reliability and security that can only be provided by the def-inition of an architecture that extends the ones from HART, while minimizing protocol overhead.

Beyond the ISO 7-layer model, WirelessHART also specifies methods to integrate devices in an industrial control system. One example is setting up sensors' operational ranges, tags, alarms, input types, etc.

An integrated WirelessHART solution provides a low-cost solution by using the HART infrastructure. Using standard HART commands, users can provision and maintain field instruments with existing configuration devices.

In addition, the wireless infrastructure provided by WirelessHART provides secure transmission to avoid either eavesdropping or alteration of the data. WirelessHART ensures that the security complies with the most stringent requirements within the process industry.

Battery Life

To increase battery life, WirelessHART specifies a degree of intelligent reporting. This goal is achieved by programming the WirelessHART-enabled device to transmit data only when commanded or when a process variable change occurs.

Furthermore, WirelessHART uses time division multiplexing, in which a device schedules its communications within the network while sleeping most of the time. For example, WirelessHART can schedule communications so that battery-operated devices can operate many years without a battery exchange.

Finally, scheduled timeslots can be used to schedule transfer of data by reporting data on exceptions (e.g., alarm and events).

The WirelessHART time division multiplexing also provides a time stamp on the information transfer of process variables (PV) to enhance the real-time value of the communications performance.

Central Network Coordination

The time division multiplexing mechanism and other WirelessHART functions are centrally coordinated. The time coordination ensures that the network can operate without collisions caused from its own devices while minimizing interference with other systems.

Using central coordination, the network manager can respond in real-time to elements affecting the performance of the network and reschedule transmissions to avoid congested mesh routes and frequencies experiencing performance degradations due to jamming, multipath fading, or interference.

Building all these features into the WirelessHART standard simplifies installation while providing an open network for multiple-vendor devices to operate within.

THE WIRELESSHART SYSTEM

WirelessHART uses IEEE Std 802.15.4 compliant radios operating in the 2.4 GHz Industrial, Scientific, and Medical (ISM) radio band. The radios employ direct-sequence spread spectrum (DSSS) technology and channel hopping for communication security and reliability, as well as TDMA-synchronized, latency-controlled communications between devices on the network (TSCH). This technology has been proven in field trials and real plant installations across a broad range of process control industries with very high levels of reliability.

Each device in the mesh network can serve as a router for messages from other devices. In other words, a device does not have to communicate directly to a gateway; the device just forwards its message to the next closest device. This practice extends the range of the network and provides redundant communication routes to increase reliability.

The network manager determines the redundant routes based on latency, throughput, efficiency, energy consumption, and reliability. To ensure the redundant routes remain open and unobstructed, messages continuously alternate between the redundant paths. Consequently, like the Internet, if a message cannot reach its destination by one path, it is automatically rerouted to follow another known-good, redundant path with no loss of data.

The mesh design also makes adding or moving devices easy. As long as a device is within range of others in the network, it can communicate.

To meet the needed flexibility for different application requirements, the WirelessHART standard supports multiple messaging modes, including:

- One-way publishing of process and control values

- Spontaneous notification by exception
- Ad-hoc request/response
- Auto-segmented block transfers of large data sets

These capabilities allow communications to be tailored to application requirements, thereby reducing power usage and overhead.

An additional capability relates to instrumentation commissioning. The WirelessHART system was designed to enable the commissioning of new instruments in a simple yet robust secure joining procedure, using existing tools from HART-based systems.

WirelessHART Coexistence

WirelessHART was designed to operate in a way that maximizes coexistence in the 2.4 GHz band. This is achieved by the use of a collection of coexisting measures:

- Channels quality assessment that builds on IEEE Std 802.15.4 Clear Channel Assessment and channel blacklisting
- Channel hopping to avoid busy channels and minimize impact of multipath fading
- Time diversity by the use of schedule transmission by TDMA
- Use of short-duration data packets (good neighbor)
- Variable transmit power
- Path diversity enabled by mesh networking

WirelessHART uses channel-hopping techniques to continually switch transmissions among different frequency channels. This practice prevents potentially disruptive radio signals from blocking WirelessHART transmissions. The direct-signal spread spectrum provided by IEEE Std 802.15.4 allows sharing the band with minimal effect on other systems because of the energy-spreading effect and the coding diversity of this modulation technique.

The media access control (MAC) header defined by WirelessHART is also used to further ensure coexistence with other IEEE Std 802.15.4-based networks such as Zigbee. The key for this element is the different frame header key: If a Zigbee device receives a message from a WirelessHART device, then it will just drop it and vice versa.

Timeslots allow periodic transfer of data and also reporting of data on exception (alarm, alert, event). Each instrument is time synchronized, and so each data packet contains a time stamp. All wireless instruments that are part of a single wireless network share access to the frequencies available within the ISM band. To achieve an efficient communications protocol, each network instrument has a common sense of time to avoid data collisions and synchronize data transmission and reception. Timely access to the network is achieved by dividing time into slots and distributing these time slots to individual instruments.

Each WirelessHART device has allocated time slots to accommodate the required data refresh rate, with the result that each network can support instruments reporting at different rates. Distribution of time slots is automatic and based upon the update rate entered into each instrument during setup.

With time synchronization, each instrument is aware of the sequence of channels used during the channel-hopping procedure ensuring that the transmitting and receiving WirelessHART devices are not only time synchronized but frequency synchronized too.

WirelessHART Network Reliability

The previous coexistence elements not only allows a harmonious operation in the 2.4 GHz band but augment the system reliability to very high levels. For example, stress tests have shown WirelessHART maintaining 99.999% data reliability in a 20+ instrument mesh. With this reliability performance, it is possible to observe the system experiencing a loss of wireless packets. However, because of the multiple redundancy mechanism, the system operates without errors.

There are three key components of the protocol that contribute to end-to-end network reliability, simple installation and power efficiency:

1. Time-synchronized communication makes the best use of communications capacity and power conservation.

2. Channel hopping reduces interference effects while mitigating eavesdropping risks.

3. Fully redundant mesh routing provides a robust and adaptable network that can dynamically account for changing conditions.

Devices designed to the WirelessHART standard will be interoperable with one another independent of the vendor and support tools that will work with both wired and WirelessHART devices.

Time Division Multiple Access (TDMA) is used to schedule transmissions over the network. A WirelessHART schedule is defined as a series of 10 ms time slots used to coordinate data transmissions. This setup results in reduced power consumption, the elimination of data collisions within the network, and synchronized transmissions to effectively use bandwidth and reduce latency. In addition, the TDMA mechanism allows the use of redundant time scheduling.

Mesh Networking

WirelessHART defines a centrally managed mesh network where each device within the network can transmit its own data as well as relay information from other devices in the network. This key element provides highly reliable end-to-end data communication. Each transmitting field device is provided with two or more routes to send data to the network gateway. These routes provide redundancy when a preestablished route is temporarily blocked either physically or by interference. A new route automatically establishes if a route is seen to be permanently blocked.

The mesh network defined by WirelessHART allows each instrument to have redundant paths back to the host system. The *network manager* automatically allocates these pathways to maintain real-time redundancy control.

Network Device Types

Figure 9–1 shows the network components specified for WirelessHART. The plant automation network can be based on a standard protocol, or it can be a proprietary network. The gateway is connected to the wireless network through its network access points. More than one network access point can be used to increase the throughput and improve the overall reliability of the wireless network.

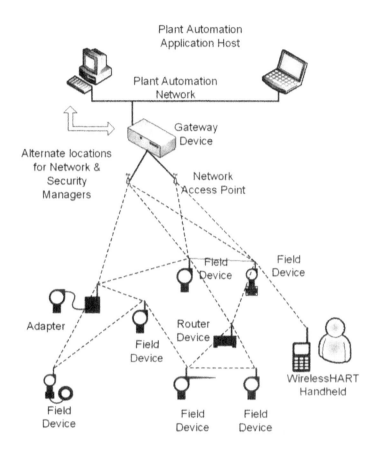

Figure 9–1:   WirelessHART network components

There are a variety of devices that can be connected to the WirelessHART network:

- Field device
- Adapter
- Router
- Gateway
- Handheld device

All WirelessHART network devices transmit and receive PPDUs and perform the basic functions necessary to support network formation and maintenance. All network devices can source and sink packets and can route packets received from other devices in the network. Each network device also has attributes holding information on update rates, sessions and device resources covering items such as the size of the superframe. Each network device contains a list of neighbor devices that it has identified during its listening operations.

Moreover, all network devices are assigned a unique identification (ID) number at the time of manufacturing. This unique ID follows the format for the IEEE Extended Unique Identifier (EUI) formed as the concatenation of a 24-bit Organizationally Unique Identifier (OUI) value assigned by the IEEE Registration Authority to the HART Communications Foundation and a 40-bit extension identifier.

Field Device

A field device is directly connected to the process to measure it, control it, or both. It is a producer and consumer of data packets capable of routing data packets received from other network devices compliant to the WirelessHART standard. The field device may be line powered, loop powered, battery powered, or powered in some other fashion.

Every device within the WirelessHART network can route packets. However, the network manager controls the routing capability of each device according to a mesh optimization algorithm and potentially user-provided constraints.

Adapter

An adapter device provides physical and logical connectivity to non-native communicating devices to the wireless network. The adapter makes the non-native device look like a WirelessHART *field device*. It uses internal routing tables to coordinate traffic flow between the wireless network and its non-native communicating subdevices. An adapter is not directly connected to the process, but it supports all commands required of the field device, including publishing process data on behalf of the non-native connected field device. In response to a *Identify* command issued by the *gateway*, it sets the Device Flag field to 0x04 (Protocol bridge device).

WirelessHART allows the implementation of a special class of HART adapters. This adapter performs as a network device that plugs into an existing HART instrument (wired) to pass the instrument data through a WirelessHART network to the host application. The HART adapter can be located anywhere along the instrument 4–20 mA cable, and it could be battery powered or obtain its power from the 4–20 mA cable.

Router

A *router* is a specialized device designed to only forward packets from one network device to another. It uses networking resources to find the next hop neighbor to which the received packet has to be forwarded.

In WirelessHART, stand-alone routers are not required because all network devices support routing. However, it could be beneficial (for example, to extend the network or to save the power of a field devices in the network) to add additional devices to improve routing in the network. A router is typically not connected to the process instrumentation.

Gateway

A *gateway* device connects the WirelessHART network to a plant automation network, allowing data to flow between the two networks. It provides host applications access to the network devices. A gateway device can be used to convert from one protocol to another, as go-between two or more networks that use the same protocol, or to convert commands and data from one format to another. The WirelessHART network also uses the *gateway* as the source for the synchronized clock used by the timeslots and superframes; WirelessHART defines the use of a single logical gateway per network.

Although only one logical gateway entity exists in a WirelessHART network, it is possible to design gateways to have multiple access stages to other networks.

The gateway functionality logically contains a *network access point* (or simply *access point*), a *network manager,* and a *security manager.* However, each of these components can be physically bundle together within the gateway device or physically distributed in different network hardware infrastructure.

The WirelessHART gateway provides a sink or source point for the wireless network traffic. The gateway also provides a connection to the network manager. The gateway is identified within the network by a fixed unique ID (0xF981000002). In addition, the *gateway* is located at the root of all network graphs, allowing PDUs to be routed to the most convenient network access point. The access point works as a network layer router, containing a unique ID and network address.

The gateway uses application layer services to communicate with network devices. Furthermore, the gateway works as a server responsible for collecting and maintaining cached data and command responses from all devices in the network. These cached responses correspond to publish data messages, event notifications, and command responses. The cached responses are returned immediately to host application requests causing a reduction of the network communication load and improving the power utilization and host application responsiveness.

Network Access Point

The *network access point,* also referred as *access point,* allows interconnection of the WirelessHART network to the gateway by providing the physical interface into the wireless field network. A *gateway* can use more than one *network access point* to improve the effective throughput and reliability of the network.

The *network manager* is responsible for configuring the routes in the *access point.* When a gateway uses multiple network access points, the *network manager* provides redundant routing for each network device.

> The communication between the gateway and any of its logical components, such as the access point, network manager, and security manager, is outside the scope of the WirelessHART standard.

Network access points propagate the clock to the field network; there is always at least one network access point performing this task. If there is more than one *network access point,* the gateway ensures that their respective clocks remain synchronized with each other, which leads to one of three possible configurations:

- A single network access point (see Figure 9–2)
- Multiple network access points, each providing the network clock (see Figure 9–3)
- Multiple network access points with at least one not providing the network clock (see Figure 9–4)

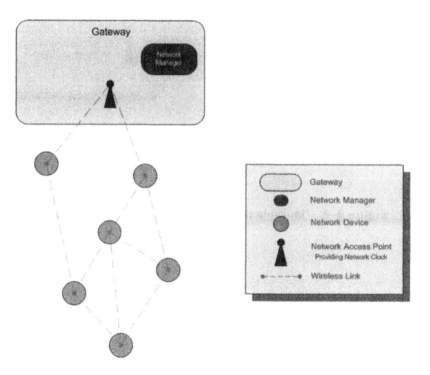

Figure 9–2:   Single network access point with clock

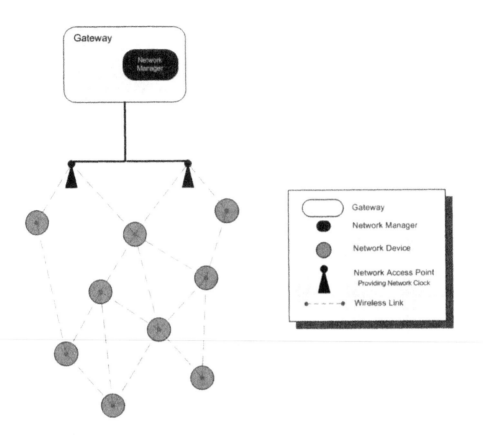

Figure 9–3:   Multiple network access points with clocks

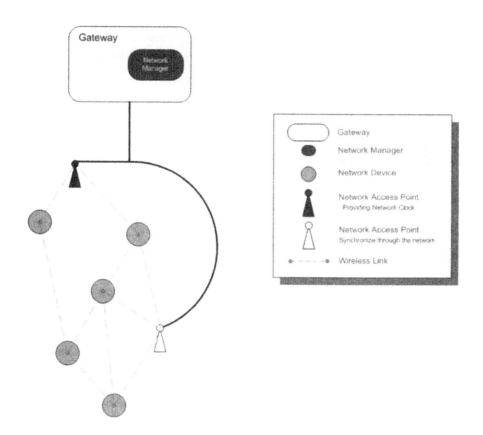

Figure 9–4: Network access point not providing network clock

Handheld Device

A handheld device is used in the installation, control, monitoring, and maintenance of a WirelessHART network. It is a portable piece of equipment normally operated plant personnel. There are three types of WirelessHART
handheld devices:

• Wired handheld device that interfaces to the existing HART network to
commission WirelessHART devices.

• Wireless connected handheld device that communicates directly to the
WirelessHART network. This handheld device joins the network as a wireless field device.

- Plant automation network–connected handheld device that connects to the plant automation network through some other networking technology. This device communicates with the network devices through the gateway device in the same way as the external plant automation host does.

Network Manager

The *network manager* does not directly connect to the wireless network, but it is logically part of the *gateway*. It has its own unique ID, and thus it communicates with other network devices using the application layer protocol of this standard. The network manager is responsible for configuring the network, scheduling communication between network devices, managing the routing tables, and monitoring and reporting the health of the wireless network. It contains a complete list of network devices and ensures that each has a network-unique short 16-bit address.

As part of its system functions, the network manager collects performance and diagnostic information. This information is accessible during run-time, making it possible to view and analyze the behavior of the overall network. If problems are detected, the reconfiguration of the network is performed while the network is operating. This network grooming is performed continuously as the overall network operation and performance varies due to changes in network load and environmental conditions.

The network manager and the gateway may be physically combined or may be separate. In addition, both establish and maintain a secure communication channel with each other. All communications between the network manager and the network devices pass through the gateway; therefore, the gateway is the entity in charge of routing PDUs to the specified network destination—network device, host application, or network manager. Chapter 10 elaborates on the operation of the WirelessHART network manager.

Security Manager

The *security manager* is responsible for the generation, storage, distribution, and management of encryption keys. In addition, it holds the list of authorized devices to join the network as defined by a configuration policy.

To accomplish its mission, WirelessHART uses three security keys: join key, network key, and session key. The network and session keys are provided to the network manager, and the join keys are provided to the network devices. These keys are used for device authentication and encryption of data in the network.

The WirelessHART network manager and devices require all three keys. The join key and the network key are used during the join sequence. Session keys are allocated during the join process and used thereafter.

WirelessHART defines one security manager associated with each wireless network. However, the security manager may be a centralized function in some plant automation networks, servicing more than just one wireless network and in some cases other networks and applications.

The functional requirements and the operation of the security manager and its communication with the network manager are outside the scope of the WirelessHART standard. Chapter 12 expands on the topic of WirelessHART security and expands coverage on the security manager.

Power Supply and Power Management

Devices complying with WirelessHART can be powered either by mains power, battery, energy-harvesting devices (for example, solar, vibration), or a combination thereof.

As will be shown later, a network manager centrally coordinates a WirelessHART network. This device is in charge of defining a time division multiplexing superframe structure with dynamic time slot allocations for each network device. The time slot scheduling is designed to sustain the communication needs of the specific industrial instrumentation. To conserve power, network devices become inactive (go to sleep) during the time where no communications is scheduled.

In addition, as part of the power management mechanism of the WirelessHART system, WirelessHART network devices report the condition of their power supply so that if they are battery powered the battery can be replaced before it is exhausted.

Electronic Device Description Language (EDDL)

Sensors and actuators management is an ongoing process, beginning with initial installation and commissioning and continuing with periodic calibration and diagnostic checks over the life cycle of a device to ensure that the device and the plant operate reliably and with high performance. To avoid mistakes in setup and maintenance, the software integrated in the control system (or separate system used for device management) typically provides a user interface that displays full device information as intended by the device manufacturer, in a consistent manner.

The Electronic Device Description Language (EDDL) is a text-based, domain-specific language that is normally used in the industrial applications to describe the characteristics of field devices. EDDL technology enables the integration of industrial product details using standardized tools for the management of the entire product life cycle.

In addition, EDDL is flexible enough to describe devices with different communication interfaces such as HART, WirelessHART, and PROFIBUS, with the higher layer interpreter, as illustrated in the example shown in Figure 9–5.

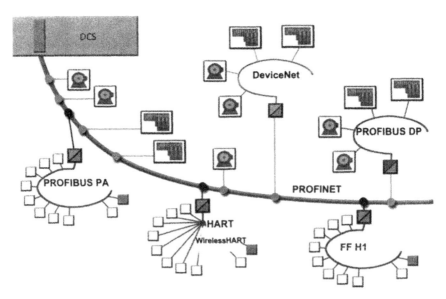

Figure 9–5:   Example of device integration by EDDL on different industrial networks

EDDL provides a standardized form and structure for host systems and hand-held communicators to access and display information in field instruments independent of the communication protocol or device operating system. EDDL defines a domain-specific language for describing the properties of industrial automation system components, such as generic digital and analog input/output modules, motion controllers, human machine interfaces, sensors, closed-loop controllers, encoders, hydraulic valves, and programmable controllers. The properties described using EDDL include:

- Device parameters and their dependencies
- Device functions, for example, simulation mode, calibration
- Graphical representations, for example, menus
- Interactions with control devices
- Graphical representations
  - Enhanced user interface
  - Graphing system
- Persistent data storage

The semantic and lexical structure of EDDL is specified in a syntax-independent manner in the standard IEC 61804-3.

From the viewpoint of the ISO/OSI model, EDDL is above Layer 7. However, the EDDL application uses the communication system to transfer its information; EDDL contains constructs that support mapping to a supporting communication system, including, of course, WirelessHART.

> EDDL also describes the management of information to be displayed to the user. However, the specific representation of such visualization is not part of the EDDL standard.

Mapping WirelessHART onto the ISO/OSI Basic Reference Model

The WirelessHART protocol was designed using the principles, methodology, and model of ISO/IEC 7498-1. The ISO/OSI model provides a layered approach to communications standards, whereby the layers can be developed and modified independently. WirelessHART specifies functionality from top to bottom of a full ISO/OSI stack including functions for the users of the stack. Functions of the intermediate ISO/OSI layers, layers 3 through 6, may be consolidated following the simplified fieldbus model shown in Figure 9–6. In this case, WirelessHART's layers 5 and 7 are logically aggregated to the fieldbus Application layer, and layers 3 and 4 are aggregated to the fieldbus data link layer.

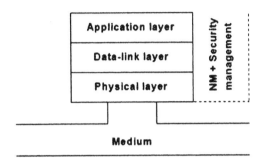

Figure 9–6: Basic fieldbus reference model

Table 9–2 shows the ISO/OSI layers, their functions, and the equivalent layers in the WirelessHART reference model.

Table 9–2: ISO/OSI and WirelessHART layers

OSI layer	Function	WirelessHART layer
7 Application	Translates demands placed on the communications stack into a form understood by the lower layers and vice versa	Command-oriented interface. Predefined data types and application procedures.
6 Presentation	Converts data to and from standardized network formats	↑
5 Session	Creates and manages dialog among lower layers	↑
4 Transport	Provides transparent reliable data transfer (end-to-end transfer across a network that may include multiple links)	Auto-Segmented transfer of large data sets, reliable stream transport. Negotiated Segment sizes ↓ or ↑
3 Network	Performs message routing	Power-Optimized Redundant Path. Mesh to the edge Network. ↓ or ↑
2 Data-link	Controls access to the communication medium and performs error detection. (point-to-point transfer on a link)	Secure & reliable ,time synched TDMA/CSMA, frequency agile with ARQ
1 Physical	Encodes and decodes signals for transmission/reception in a form appropriate to the communications medium. Specifies communication media characteristics	2.4GHz Wireless, 802.15.4-based radios, 10dBm Tx Power

NOTE ↓ and ↑ indicate that the functionality of this layer, when present, may be included in the fieldbus layer that is nearest in the direction of the arrow. Thus network and transport functionality may be included in either the data-link or application layers, while session and presentation functionality may be included in the application layer but not in the data-link layer.

WirelessHART Network Planning and Installation

In general, the installation of any wireless network requires careful planning to ensure acceptable levels of performance and reliability. This is the case for both wired and wireless networks. WirelessHART was designed to minimize the planning and installation process by engineering capabilities for self-configuration and self-healing, commanded by the network manager.

Devices wanting to form a WirelessHART network require only a join key and a network ID. However, the network manager may allow operators to pose constrains in the network formation process by manipulating parameters such as:

- Topology control for specialty applications
- Network size control
- Propagation latency control
- Network device refresh dates
- Rules for the provisioning of redundant data paths
- Channel blacklisting
- Manipulation of routing capabilities
- Handling of multiple-cast options among multiple access points

The optimizing algorithm within the network manager is not part of the WirelessHART standard. Different implementations can provide either a simple configuration interface, hiding the complexity of the network inner-working or a more involved control of the network., exposing elements.

Chapter 10 The WirelessHART Network

As highlighted in the previous chapter, the WirelessHART *gateway* provides the connection to a host network such as PROFIBUS and WirelessHART field devices. The gateway can also host the network manager and security manager, although these functions can also exist at the host application level.

The network formation process is initiated with the network manager. This device contains the necessary algorithms to build and maintain the mesh network. The network manager analyzes in real time, the quality of service information coming from the WirelessHART PDUs to identify the best paths, schedules (distribution of 10 ms time slots), frequency channels, and other parameters, to maintain a predetermined level of performance.

> The WirelessHART standard does not impose any restrictions on how the security manager, network manager, and network access points are implemented. Early implementations of WirelessHART systems have the network manager, security manager, and network access points built into the gateway hardware. However, this architectural option is vendor specific. Next generation products are now separating the network access point, gateway, and network manager.

Every field device in the WirelessHART network has the capability to route messages. However, the network manager controls the actual routing functionality. For particular applications, the network manager can disable routing for specific field devices according to any specific criteria. However, manual disabling of routing is not recommended because it makes the network management function suboptimal. For example, it might lower power consumption for some devices but increase consumption in others or create other side effects such as adding congestion points across the mesh network. The network manager running its optimizing processes already considers battery life maximization and traffic management optimization, among other quality of service parameters.

NETWORK SELF-HEALING

WirelessHART networks allow the use of dedicated routing devices, or *routers*. These devices are not connected to the industrial process but help extend the network coverage range. In addition, these routers help increase the robustness of existing operational networks, e.g., go around an existing or new obstacle.

The built-in capability of every WirelessHART Network Device to route messages under the command of the network manager simplifies network planning and deployment. In addition, this capability is the key functional element the self-healing capability of the wireless mesh created by a WirelessHART system.

NETWORK MAINTENANCE

The network manager is in charge of scheduling time slots, managing dynamic channel allocation, and determining the paths (optimal and redundant) to maintain a specified level of quality of service. Network maintenance activities involve discovering potential neighbors, gathering statistics about the communication channel to each existing neighbor, and maintaining time synchronization with each of them. To augment reliability, each *network device* in a WirelessHART network maintains a list of neighboring devices with which it can communicate directly. To cope with the dynamics of the RF environment—propagation effects and multipath fading—the *network device* keeps its list of neighbors current.

Two special data link PDUs (DLPDU) assist in building and maintaining a devices list of neighbors. These are Advertise DLPDU and Keep-alive DLPDU. These DLPDUs contain sufficient information for new neighbors to be discovered or for a newly installed device to request admission to the network. If accepted, the new device can become a neighbor of the advertising device. As part of its own network management algorithm, the network manager can schedule transmission of these DLPDUs.

Every successful communication with a neighbor confirms the neighbor's presence and allows the quality of the communication link to be assessed. Since communications only occur when the device has a DLPDU for the neighbor, there can be long intervals during which the link is not exercised. Keep-alive DLPDUs are used to probe quiescent links and to maintain time synchronization.

WIRELESSHART NETWORK TOPOLOGY

WirelessHART networks can be configured in a number of different topologies to support various application requirements including the following ways:

- *Star network*: Star networks (see Figure 10–1) have just one router device that communicates with several end devices. The router device can also be the *network access point* of the *gateway*. This is one of the simplest network topologies. A star network might be appropriate for small, low-coverage applications. Chapters 3, 5, and 6 cover the topic of star topology networks.

- *Mesh network*: Mesh networks (see Figure 10–2) are formed by network devices that are all router devices. Mesh networks provide a robust network with redundant data paths that can adapt to changing RF environments. Chapters 5 and 6 cover in detail the subject of mesh networks.

- *Star mesh network*: Star mesh networks (see Figure 10–3) are a combination of the star network and mesh network. In essence, this topology is just another type of mesh.

Figure 10–4 shows a possible redundant architecture.

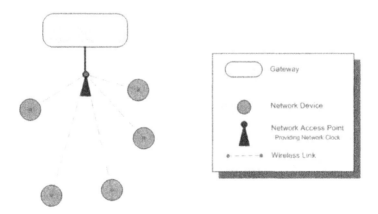

Figure 10–1:   Sample topology of a WirelessHART star network

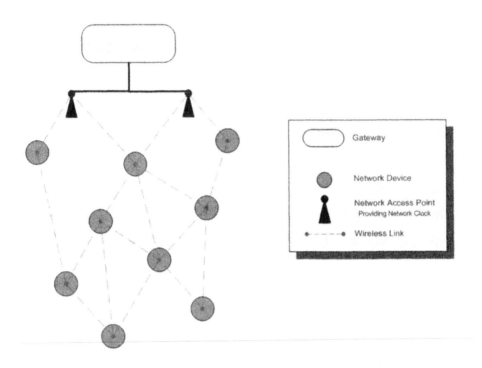

Figure 10–2: Sample topology of a WirelessHART mesh network

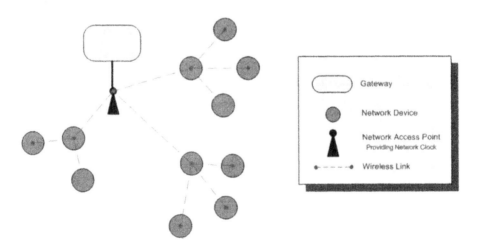

Figure 10–3: Sample topology of a WirelessHART star-mesh network

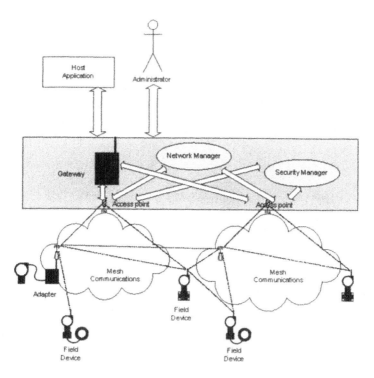

Figure 10–4:   Network architecture in wireless network

The mesh topology of a network provides the most robust topology for a wireless network because there are multiple redundant pathways to move data from source to destination. Mesh topology provides high data reliability, which is a key network requirement for the process automation industry. Figure 10–5 illustrates a typical WirelessHART network composed of multiple network devices integrated with actual industrial field instrumentation.

The *network manager* builds and maintains the mesh network. It identifies the best paths and manages distribution of slot time access (WirelessHART divides each second into 10 ms slots). Slot access depends upon the required process value refresh rate and other access (alarm reporting—configuration changes).

If a network path becomes unreliable or broken due to an obstruction, then the network manager automatically finds an alternate route for the information.

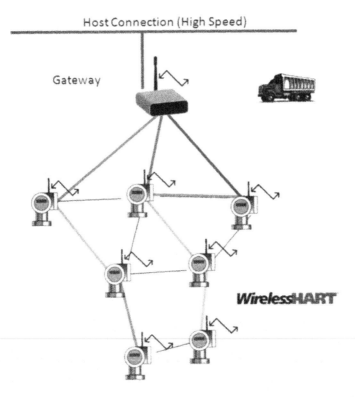

Figure 10–5: Mesh topology of a wireless network

NETWORK MANAGER

The *network manager* is responsible for the overall management, scheduling, and optimization of the WirelessHART network. As part of its tasks, the *network manager* initializes, maintains, and manage network communication parameters. The network manager provides mechanisms for devices to join and leave the network, and manages dedicated and shared network resources.

The *network manager* is an application that coordinates the mesh network and associated *network devices* in a manner that maintains a quality of service demanded by the industrial application. The network manager performs the following functions:

- Forms a star or mesh network
- Allows new devices to connect to the network

- Determines optimal communication schedule of *network devices*
- Establishes the redundant data paths for all communications
- Communicates with the security manager to obtain encryption and authentication data to secure the network
- Monitors the network performance in real time and makes the necessary changes to sustain a preestablished quality of service

The mesh network architecture does not restrict where the location of the network manager in the plant automation network. The network manager may be physically located on a gateway, network device, or on another system device. The network manager and the gateway work as a single logical entity, so that it is most likely that network manager implementations collocate these device with the gateway in the same physical box. However, nothing precludes the network manager implementations in a completely separate physical box. In addition, it is possible to have redundant network managers connected to one network, but there shall be one and only one active network manager per instance of a wireless network.

There is one network manager per mesh network (and subnetworks). Figure 10–6 shows a diagram of the network manager relationships with the rest of the WirelessHART network.

The network manager has a fixed, well-known unique address set to 0xF980000001; (16-bit address = 0xF980), which is used as destination for the application layer PDUs from field device to the network manager.

The network manager is also responsible for collecting and maintaining diagnostics about the overall health of the network. These diagnostics are available to be reported to host-based applications. The diagnostics are also used to adapt the overall network to changing radio-frequency conditions to satisfy quality of service demands from the application.

For the network manager to perform its complete set of functions, it needs information about the network devices, information about how the network is to be used, and feedback from the network on how well the network is performing. Configuration and setup information about network devices is read directly by the network manager. Communication resources are requested by network devices, host applications, and plant users. Feedback on how well the network is performing is provided by the devices through health reports and diagnostics built-in the WirelessHART protocol.

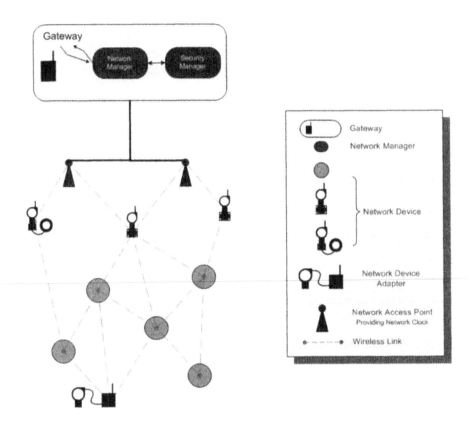

Figure 10–6:   Network manager in a WirelessHART network

In this case, the network manager monitors the mesh network in real time for any performance changes. The network manager collects the following performance information:

- Path statistics
- End-to-end reliability
- Signal strength
- Battery life
- Communication status

The network and path performance is used to repair data paths if one becomes blocked or to find redundant ones in case of communication issues. The optimization algorithms used within the network manager enables the self-healing attributes of the WirelessHART technology, allowing it to deliver the robust data communications required by end users in the industrial space. This automatic self-healing process reduces engineering input, simplifying the use of WirelessHART.

The interface between a gateway and the network manager is outside the scope of this standard. However, the network manager and the gateway are responsible for establishing a secure connection with each other, and maintaining this connection to carry control and data traffic. Thus, it is not necessary for the gateway to go through the normal network device join process. After the gateway connects to the network manager, the network manager may configure the gateway to begin the network formation process.

Network Formation

A key feature of a WirelessHART network is its ability to self-organize. The *network manager* orchestrates this feature with four key elements:

- Advertising
- Joining
- Scheduling
- Discovery

Advertising

As part of advertising, devices that are already part of the network may send packets announcing the presence of the network. Advertisement packets include time synchronization information and a unique network identification. Devices trying to join the network listen for these packets and try to match the advertised network ID with their own. When at least one advertisement is heard, the new device can attempt to join the network.

Joining

A new *network device* joins the network by sending a join request packet encrypted with the join key with a payload containing the field instrument unique ID, through an advertising node to the *network manager* and *security manager*. If it is authenticated, then the *network manager* will respond with an activation packet, accepting the new device into the network and setting up links between the new device and other existing nodes. In addition, after authentication, the network manager makes the appropriate time slot allocation, and routs information to the *network device*.

Figure 10–7 shows an example of a message sequence chart representing a *network device* joining a WirelessHART network under a simplistic mesh network topology. The example illustrates the details of the joining process.

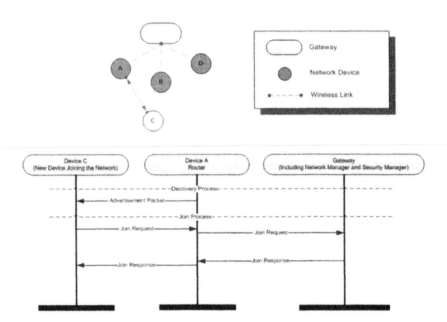

Figure 10–7:   Message sequence chart of a network device joining a WirelessHART network

Scheduling

After a new *network device* joins the network and has initial communication resources allocated (session, superframes, links, etc.), it then requests additional network resources based on its configured publish rate. The *network manager* can allow or deny these network resources. If the network manager allows the request, it then sends out a schedule to the new device and any intermediate routing devices with the new schedule information. A new net-work starts when a gateway device (typically at the network manager's request) starts to advertise. Being the first device in the network, the gateway starts its own schedule, to which other nodes later synchronize. Schedules allow network devices to report or publish process variables on a periodic basis or enable commands from the gateway to reach these devices. Figure 10–8 shows an example of a message sequence chart of a network device publishing process information. Similarly, Figure 10–9 presents another example message sequence chart illustrating a typical command-response message sequence chart within the WirelessHART protocol.

Discovery

One other aspect of WirelessHART's self-organization is the discovery function using dedicated discovery slots. This feature finds other poten-tial neighbors that can be used to form other redundant paths.

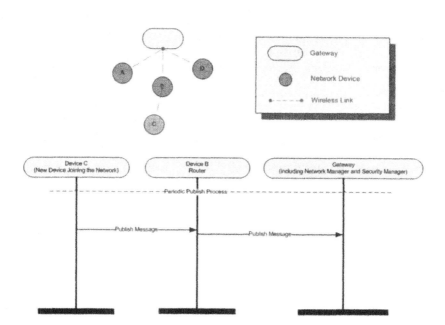

Figure 10–8:   Message sequence chart of a network device publishing process control information in a WirelessHART network

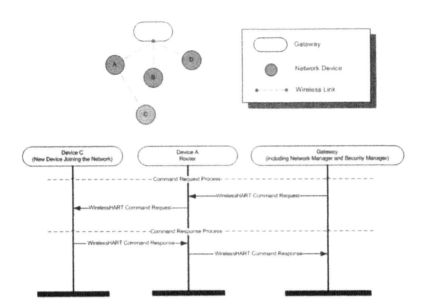

Figure 10–9: WirelessHART command/response message sequence chart

NETWORK MANAGER FUNCTIONS

The network manager performs the set of functions summarized in Table 10–1.

Smart Updates

The network manager can automatically alter its data publish rate based upon process conditions. This ability is called Smart Updates. This mechanism augments reliability in the event of a varying RF environment, maximizes battery life, and allows exceptional communications to go through. that is, the process variables and status being reported without a polling action by host application (through the gateway).

Figure 10–10 shows an example of Smart Updates, triggered by a threshold value of a process variable. In this example, when the process variable is below a certain level, the value needs to be sampled (refreshed) and communicated with a sampling rate of 20 seconds. If the variable crosses the user-defined threshold, a time stamped alarmed is generated and the sampling (and communications) rate is increased to once every 10 seconds.

Table 10–1:   Network manager requirements (Sheet 1 of 3)

Network function	Requirement
Network formation and configuration	Provide logic for initializing itself and starting up the network.
	Manage topology adapt the network to changes as diagnostic information is reported from devices.
	Manage distribution of the network key that is provided by the security manager to all network devices. Separate network manager and gateway keys are used for unicast and broadcast traffic that originates from each of them.
	Manage join process. The network manager validates the device that tries to join the network. After authenticating a network device, the network manager shall give the joining network device the network key and four session keys: 1- network manager unicast session keys 2 - network manager broadcast session key 3 - gateway unicast session key 4 - gateway broadcast session key Device shall be configured with network_ID before it attempts to join the network, so that it can find the right network to join.
	Assign 16-bit address nickname. The network manager shall assign and manage network unique 16-bit address (network addresses) to each network device. The network manager is responsible for ensuring that the Neighbor table inside each device is up-to-date.
	Establish a connection with the gateway. Whenever the gateway receives a message destined for the network manager gateway forwards it to the network manager. The network manger shall be connected to the network via the gateway.
	Configure at least one of the network access points connected to the gateway to provide the network clock.
	Manage network configuration. Maintain a full map of the network configuration, including any information about the network that has been distributed to network devices.
	Respond to requests for network information. For example when a host application makes a network management information request for a device in the network, the network manager is responsible for providing the response.

Table 10–1:   Network manager requirements (continued) (Sheet 2 of 3)

Network function	Requirement
Routing	Create and manage network route. The network route represents a complete map of the network.
	Manage neighbor tables. The network manager collects network statistics and neighbor table information from each device through periodic health reports. This information is used to adapt the network to changes.
	Build route tables for graph routing. Graph routing is ideal for both scheduled upstream and downstream communications. Upstream communications include process measurements and alarms. Downstream communications include set point changes to actuators.
	Build source route lists for source routing.
	Allocate communication resources to the network devices, gateway, and to itself so that it has the network capacity allocated to manage the network and the devices have sufficient network capacity to communicate.
Network schedule	Create superframes. Multiple superframes are used to support communications at specific scan rates. Additional superframes shall be allocated to support device management and diagnostic applications which require large amounts of traffic for short periods of time.
	Assign links in superframes.
	Create link tables. Each link includes exactly one slot associated with a superframe, its type (normal, advertising, discovery), its options (transmit, receive, shared), neighbor information, channel offset, and the device connected to this link.
	Activate and deactivate superframes in response to application demands.
Channel management	Keep track of blacklisted channels. Note: Assigning a channel to the blacklist is a manual operation.
	Provide channel offset. The channel offset is used to calculate the channel number when channel hopping. The channel offset takes on a value of 0 to (number of channels minus number of blacklisted channels).

Table 10–1:   Network manager requirements (continued) (Sheet 3 of 3)

Network function	Requirement
Network diagnostics and adapting	Maintain overall wireless network diagnostic information. If a network device has not received a packet from one of its neighbors within the Keep_alive_time, the device sends a path-down notification to the network manager indicating that the path is no longer available.
	Maintain record of health information about each network device.
	Allocate communication resources as requested by network devices. Devices request network capacity to support data publication, event notification, and block mode traffic. Gateways request network capacity to support client demands. Network traffic is biased towards a particular path by increasing or decreasing the number of links through a particular device.
	Optimize routes and schedules to improve operation of the network while conserving power within devices.

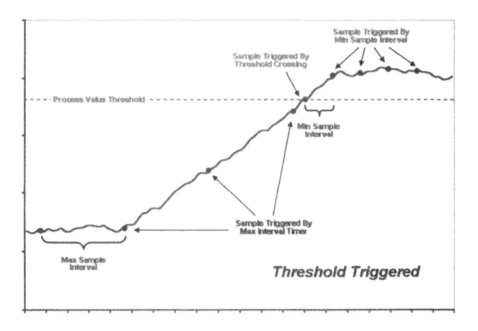

Figure 10–10:   Smart Updates: Threshold triggered

The example in Figure 10–10 uses a trigger when the process variable goes from a low to high transition across a threshold. Smart updates can also be set for high to low transitions and through a window-triggered mechanism. in this case, the process variable is reported (published) on one of two possible triggering events: sampling time and process variable value change. Figure 10–11 shows the window-triggered mechanism.

Notice that the sampling at the device level will still occur at the required update rate dictated by the process, but as long as the trigger is not reached, not all samples are communicated.

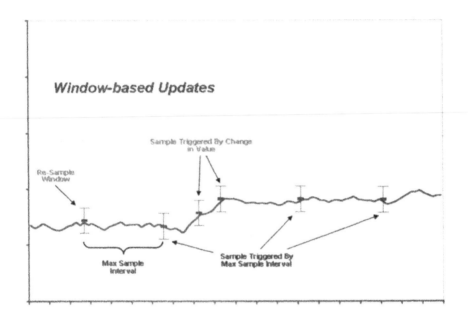

Figure 10–11: Smart Updates: Window triggered

WirelessHART also defines a mechanism to allow acyclic access to the wireless media. This access is achieved by special time slots held aside—called *shared slots*— by the *network manager*. If a device has an exceptional value to report (e.g., alarm), it waits for the next available shared slot to transmit the data.

To transfer large amounts of process information, WirelessHART specifies an enhanced burst mode efficiently publish process data. The burst messages are used to send data on a periodic schedule or on an exception basis. A burst message consists of the response packet for the specified command. The burst mode rate is configurable, based on application needs.

For control applications, the burst mode rate is determined by control loop or sequence execution requirements (e.g., the process time constant). In some cases, more than one host application will subscribe to the same published data. In these cases, the update rate is determined by the fastest requested data rate.

Chapter 11 The WirelessHART Physical Layer and Data Link Layer

The WirelessHART Physical Layer leverages the 2.450 MHz DSSS physical layer employing offset quadrature phase shift-keying (O-QPSK) modulation according to IEEE Std 802.15.4-2006. The WirelessHART physical layer is responsible for the signaling method, signal strength control, device sensitivity, and managing RF parameters for sending bits across the wireless media. To ensure conformance to international regulations, WirelessHART specifies an RF output power of 10 dBm.

The WirelessHART physical layer uses the following functionality from IEEE Std 802.15.4:

- *Data Services*: Send and receive data, manage digital information over the radio-frequency media.

- *Management Services*: Manage local parameters to control the operation of the physical layer, including management of radio operations.

Similarly the WirelessHART Data Link layer, uses the IEEE Std 802.15.4 Medium Access Control. However, to reach a level of low energy consumption, WirelessHART only makes use of the non–beacon mode complemented with a centrally controlled time synchronized superframe scheduler.

The WirelessHART standard uses IEEE Standard 802.15.4 mandatory functions, with a subset of the optional ones. Functions used the WirelessHART data link layer expand and complement the IEEE Std 802.15.4 MAC in a way that maintains full compliance to this standard. Any WirelessHART standard implementation can make use of the variety of existing off-the-shelf IEEE Std 802.15.4 devices (radio semiconductor and MAC software) existing in the marketplace.

WIRELESSHART SUPERFRAME

The stringent timing and meshing requirements from industrial process automation networks such as WirelessHART do not allow the use of the optional
GTS mechanism provided by IEEE Std 802.15.4.

A WirelessHART superframe is defined as a collection of time slots repeating
at a constant rate. Different from the optional native beacon infrastructure
provided by IEEE Std 802.15.4, WirelessHART establishes its superframe
infrastructrure without using beacons. A central network coordination function controlled by the *network manager* achieves this functionality.

One time slot can transfer the required process automation data from a single
device. Time slots defined within the WirelessHART standard have a fix duration of 10 ms.

Figure 11–1 presents an example of a WirelessHART superframe structure.

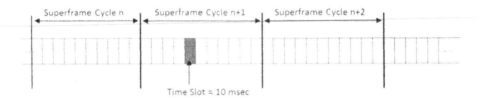

Figure 11–1:   WirelessHART superframe structure

WirelessHART uses time division to make optimum use of radio time and
bandwidth. Not all slots in a WirelessHART superframe are assigned or
active. The *network manager* is in charge of allocating time slots across all
devices in the network in a way that maximizes the quality of service required
by the application, e.g., enabling time diversity by scheduling multiple transmissions through redundand path at multiple frequency channels.

> PANs that do not want to use the built-in IEEE Std 802.15.4 superframe struc-
> ture (referred to as a non–beacon-enabled PAN) need to set both *macBea-
> conOrder* and *macSuperframeOrder* to 15. In this case, a coordinator will not
> transmit beacons, unless it receives a beacon request command. All transmis-
> sions, with the exception of acknowledgment frames and any data frame that
> quickly follows the acknowledgment of a data request command, make use of
> the unslotted CSMA-CA mechanism to access the channel.

The optimization algorithm inside the *network manager* seek to avoid RF collisions within one WirelessHART network as well as maintaining other quality of services parameters such as throughput and latency.

This optimization algorithm is outside the scope of WirelessHART standard. Vendors are free to create their own optimal network managers fit to specific industrial applications.

When *network devices* join the network, they request network resources based on their configured publish rate. The *network manager* uses this information to allocate and schedule time slots to ensure these devices can access the network with the needed bandwidth to accomplish its task.

Even when time slots are 10 ms long, WirelessHART allows the definition of *bulk transfers* or the transmission or a large block of data from the network devices. *Bulk transfers* are defined by segmenting data into data packets that can be distributed through multiple allocated time slots. A typical application for bulk transfers in the process automation industry occurs with vibration or analytical chemistry sensors.

TIME DIVISION MULTIPLE ACCESS (TDMA)

WirelessHART uses time division multiple access (TDMA) mechanism, that is, it divides time into regular time intervals to enable coordinated communications. This time division occurs at each of the IEEE Std 802.15.4 operational channels.

With the TDMA mechanism, WirelessHART can provide orderly, collision free, and deterministic transmissions required in the industrial environment. The TDMA slot allocation and overall management system is centrally managed by the *network manager*. This device hides the complexity of real-time time slot management from the end user application while enforcing deterministic industrial process monitoring and control.

Typically two devices are assigned a time slot—one to transmit a data packet and another to receive it; only one data transaction on each channel is allowed in a 10 ms time slot. A single time slot can handle only a single data packet from a network device, however within a single data packet can transmit up to eight process values.

WirelessHART defines two types of time slots:

- Dedicated time slots used for process variable reporting
- Shared time slot used for transmitting special data packets such as alarms, alerts, events, or in special optimizing cases, transmit regular process variables

At each data exchange, a time synchronization takes place at the data-link level using a common time base determined by the *network manager.*

If multiple WirelessHART networks are in range, the *network managers* of both networks should coordinate the coexistence. This mechanism is not standardized within WirelessHART, but it can be easily implemented by coordination of time slot inventories between network managers involved.

Figure 11–2 shows how one time slot is built from the perspective of a transmitting network device. Clear channel assessment (CCA) is performed at the beginning of the 10 ms time slot. CCA is followed by the actual packet transmission and the reception of the corresponding acknowlegement (ACK).

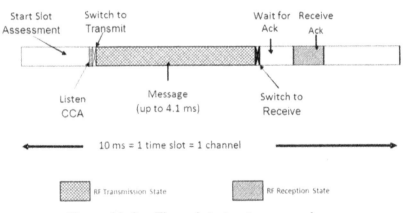

Figure 11–2: Time slot structure overview

In WirelessHART, only broadcast, advertisements, and discovery messages are not acknowledged. Time synchronization occurs either when receiving a packet or an acknowledgement from a time source. Time synchronization usually relies on acknowledged transmissions.

Figure 11–3 shows one time slot and provides an overview on the timings of a single transmission transaction. Figure 11–3 (a) shows the operation at the source network device, and the Figure 11–3 (b) shows the operation at the destination (receiving) network device. The details of slot time parameters along with the associated range of values are shown in Table 11–1. All WirelessHART packet transactions are required to comply strictly with the limits of these timing parameters, indirectly implying a certain accuracy of clock crystal much higher than the ones defined by the IEEE Std 802.15.4 standard.

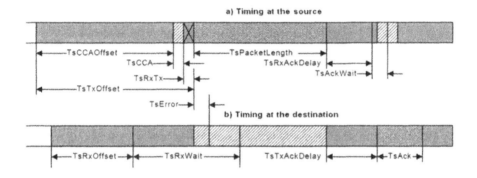

Figure 11–3:   Slot timing details

Table 11–1:   Slot timing definitions and values

Symbol	Definition	Required value
TsTxOffset	Start of the slot to start of preamble transmission	2 120 µs ±100 µs
TsRxOffset	Start of the slot to when transceiver is listening	1 120 µs ±100 µs
TsRxWait	The minimum time to wait for start of message, which correlates to the amount of drift between the neighbors that can be tolerated to maintain communications	2 200 µs ±100 µs
TsMaxPacket	The amount of time it takes to transmit the longest possible packet that includes PHY preamble, delimiter, length, and DLPDU	4 256 µs

Table 11–1:   Slot timing definitions and values (continued)

Symbol	Definition	Required value
TsTxAckDelay	End of message to start of ACK; the destination device validates the STX and generates an ACK, if required, during this interval	1 000 μs ±100 μs
TsRxAckDelay	End of transmit PhPDU to when transceiver is listening for ACK	800 μs ±100 μs
TsAckWait	The minimum time to wait for the start of an ACK	400 μs ±100 μs
TsAck	Time to transmit an ACK (26 octets) NOTE TsAck may vary based on the addressing used (16-bit or 64-bit). As an Ack may also have a length of 25, 31, or 37 octets (including preambles) that will result in 800μs, 992μs, or 1184μs.	832 μs
TsCCAOffset	Start of slot to beginning of CCA	1 800 μs ±100 μs
TsCCA	Time to perform CCA (8 symbols)	128 μs
TsRxTx	The length of time it takes to switch from reception to transmission or vice versa (12 symbols)	192 μs
TsError	The difference between the actual start of message and the ideal start of message time as perceived by the receiving device, that is, how much the receiving device perceives the transmitting device to be out of sync	

Notice that the slot start time at the source and destination network devices can be different due to different amounts of clock drift. WirelessHART compensates for these clock drifts at every data exchange with a time source, when receiving a packet from the time source by the marked arrival of the start of the packet in relation to the perceived start or when transmitting a packet to the time source by receiving a time adjustment value in the acknowledgement (calculated by the time source in relation to the arrival of the start of the packet).

When a *network device* is scheduled to receive information in a predetermined time slot, that device enters receive mode. The device starts to listen to the medium after TsRxOffset delay from the start of the slot. The receive window specified by TsRxWait permits device time to drift, while still allowing devices to communicate and resynchronize their slot timers.

TRANSMISSION OF A PACKET

On transmission, a network device waits for TsCCAOffset duration after the start of a slot to perform a clear channel assessment (CCA) with a TsCCA duration. If the channel is occupied, the transaction attempt is rescheduled for a later time slot. Otherwise the device will switch the transceiver from reception to transmission to deliver the PPDU. After that, the device begins its transmission such that the start of message (SOM) occurs exactly TsTxOffset duration from the start of the slot.

Depending on the type of WirelessHART packet, the PPDU might require acknowledged communications. If so, an acknowledgement waiting period (TsACKWait) and acknowledgement reception follow the PPDU transmission.

RECEPTION OF A PACKET

On reception, the network device enters the receive mode to listen to the medium after TsRxOffset from the start of the slot. The network device continually checks for SOMs. If SOMs are not detected within a (TsRxOffset + TsRxWait) duration from the start of the slot, then it stops listening. If the SOM is detected, then it receives and validates the received packet.

Time is allocated towards the end of the time slot for processing the received packet and preparing for the next slot (for example, assessing and prioritizing the packets queued in each *network device*). If one of the neighboring network devices acted as a time source, then the end of the slot time becomes aligned after successful communication. To do so, the *network devices* capture the end of reception of the physical layer delimiter start time and calculate the time error, TsError as the difference between the expected start time and the actual start time of the physical layer delimiter. The devices can then adjust for time drift and stay synchronized with the network.

ACKNOWLEDGEMENT TRANSACTION

Most communication packets in a WirelessHART network consist of source devices propagating a message by transmitting a packet and the destination device acknowledging the reception of that packet. For acknowledged communication, the source and destination address in the data link layer protocol data unit must contain individual device addresses.

If a received packet is valid, then the destination device inspects the destination address in the data link layer PDU (DPDU). If the destination address is not a broadcast address and there are no reception errors, the device switches from reception to transmission and initiates the transmission of the PPDU containing the ACK PPDU. The SOM of this PPDU occurs exactly at a TsTxAckDelay after the end of the transmitted source devices packet.

Meanwhile, the source device switches from transmission to reception. After TsRxAckDelay duration from the end of its transmission, the source device listens to the medium to check for ACK packet SOM for a duration of TsAckWait.

BROADCAST TRANSACTION

WirelessHART provides the capability to broadcast communications. Broadcast DPDUs include a source address corresponding to the transmitter of the message and the destination address set to the broadcast address (set to all zeroes).

On reception of a broadcast DPDU, a *network device* validates the packet and inspects the destination address. If the destination address corresponds to the broadcast address, then the receiving device will not acknowledge the transmission.

TIME KEEPING

Time synchronization across the network is essential for TDMA communications. However, regardless of the choice of device hardware time source (e.g., crystals, ceramic resonators, etc.), some skew between *network devices* will occur in practice because of temperature, voltage variations, or aging.

Consequently, the WirelessHART data-link layer has several mechanisms to synchronize the time base network-wide.

When a device receives a DPDU destined for it, its time of arrival is noted. Using this information the destination calculates the difference between the actual time of arrival and the expected time of arrival for that DPDU. This Δt is communicated in every ACK reply DPDU sent to the source device. Thus, every acknowledged transaction measures the alignment of network time between the devices.

Within the neighbor list for a given *network device*, selected neighbors specified by the *network manager* are used as time synchronization sources. When a DPDU from a time synchronization neighbor is received, the network time of the receiving device is adjusted. Time synchronization is based on either the DPDU arrival time or the Δt in an ACK DPDU, depending on which device initiated the transaction.

Depending on the time drift behavior of particular implementations of network devices, such systems will require to adjust the frequency in which they broadcast Keep-alive DPDUs; when the devices time source drifts, the device transmits Keep-alive DPDUs as needed, to its time-synchronization neighbors to stay synchronized. However, device should be built so that it does not require a Keep-alive DPDU more often than once per 30 seconds while temperature is varying 2° C per minute or less. Furthermore, device designs should tolerate one retry in case of packet loss, which corresponds approximately to a compensated clock accuracy of 10 parts per million or better. Notice that this precision is almost four times better than the minimum crystal tolerance specified in IEEE Std 802.15.4.

CHANNEL HOPPING

To further enhance reliability, channel hopping is combined with TDMA. Figure 11–3 illustrates the transmission of a communications message from a network device to the gateway using channel hopping and TDMA.

Channel hopping provides frequency diversity, which helps mitigate the effects of in-band interference and reduces multiple-path fading effects. By using channel hopping with TDMA, WirelessHART networks can offer the high levels of reliability required by the industrial environment.

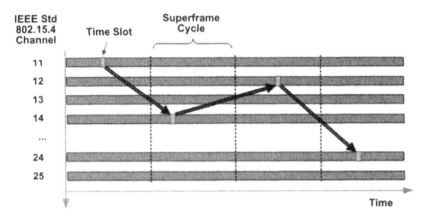

Figure 11–4:   WirelessHART channel-hopping mechanism

Communicating devices are assigned a time slot and a channel offset in a superframe specified by the network manager. This trio forms a communications link between two communicating devices. All devices can support multiple links; the number of possible links is typically equal to the number of channels used by the network times the number of time slots in the superframe. For example, using 15 channels and 9000 time slots per superframe results in 135,000 possible links.

Channel hopping provides channel diversity. So each slot is used on multiple channels at the same time by different nodes. This is achieved by creating links on the same time slot, but with different channel offsets. Each device maintains a list of channels in use and the specification for that channel. All devices in a network have identical channel lists determined by the network manager. The network manager is in charge of the dynamic assignment of links across devices in the network.

WirelessHART also supports *channel blacklisting,* that is, allowing the network manager to restrict the channel hopping on specific network devices to selected channels in the available RF band. For example, network administrators can instruct the network manager to blacklist certain channels to protect a wireless service that uses a fixed portion of the RF band that could have been shared by the devices in this network. In practice, communication in the available bandwidth by the devices using other protocols is random and uses a tiny amount of the total available bandwidth. Consequently, blacklisting seldom provides tangible benefits.

Chapter 12 WirelessHART Security

As highlighted in earlier chapters, security is a major concern in wireless communication systems. In WirelessHART, communications security has been addressed at several levels within the WirelessHART stack.

The WirelessHART technology is designed for secure industrial wireless sensor network communications while ensuring ease-of-use. WirelessHART ensures secure wireless sensor network communications by multiple features:

- Security is built in and cannot be disabled; security is always on

- Standard AES-128 bit encryption

- Key management mechanism

- End-to-end security; only the final recipient device can decrypt and use the data payload created by a source device

- Denial of Service attacks are mitigated by the use of frequency diversity caused by channel hopping

- Spoofing is eliminated by the use of message integrity checks and the maintenance of a *white list* of associated *network devices*

- Replay attacks are handled by the built-in IEEE Std 802.15.4 security engine (message freshness) and the use of TDMA technology.

Figure 12–1 shows the security mechanisms implemented by WirelessHART.

The WirelessHART security architecture was designed to help industrial plant operators to minimize, control, and audit network access while requiring very high levels of technical expertise to subvert the network. In addition, the WirelessHART security model was defined in a manner that reduces the consequences (span and duration) of any individual security breach.

To prevent eavesdropping, the security manager provides the infrastructure for the network to perform a secure join procedure, device authentication, and message encryption. Joining devices are authenticated based upon the join key and network ID. In addition, the security manager can implement a white list of expected devices to support the authentication process, as well as a black list of devices forbidden to join the network. This last feature discourages decommissioned devices to be used as cloned versions to join the network.

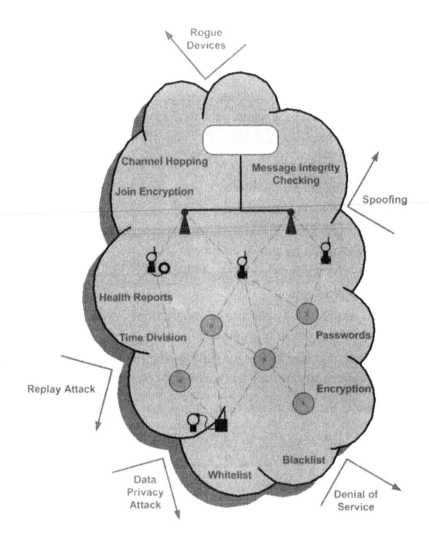

Figure 12–1: WirelessHART security features

As part of the normal network monitoring of anomalous activities, the security manager maintains statistics of message integrity code failures, authentication failures, and failed join attempts. The security manager can generate alerts based on these statistics and report showing the various states of network activity.

The security manager monitors the following anomalous network activities.

- Failed join attempts
- Excessive retries
- Message integrity code (MIC) failures
- Authentication failures

Network devices are *not* authorized to be network key servers or wireless network managers, minimizing the opportunities to design handheld hacking tools. In addition, WirelessHART does not use the TCP/IP communications stack. Therefore, it is explicitly safe from many commonplace hacking tools and hacker attacks.

DATA PROTECTION AND CONFIDENTIALITY

Data protection and confidentiality deals with maintaining the privacy and integrity of the information that is being passed over the network.

A WirelessHART sensor network provides end-to-end CCM* mode AES-128 bit encryption at the network/transport layer from the data source to the data consumer. In addition to the individual session keys, all the devices on a network share a common network encryption key to facilitate broadcast activity as needed. Encryption keys can be rotated as dictated by plant security policy to provide an even higher level of protection.

Data protection is achieved by the message integrity check feature. This function checks that data sent over the wireless network has not been altered by analyzing the built-in message integrity code (MIC) defined in each packet. When a *network device* receives a packet, it checks the MIC to confirm that the contents of the packet have not been altered while inroute by any external agents. In addition to protecting the data payload of a wireless PPDU, the MIC also protects the network routing information, preventing attacks that attempt to change the packet's network/transport layer information.

In summary, WirelessHART has one shared network key which is used for the MIC generation at the data link layer. Broadcasts have shared broadcast keys for each session used for encryption and authentication of the payload.

NETWORK PROTECTION AND AVAILABILITY

Network protection and availability, deals with maintaining the operation of the network in the face of internal or external attacks (intentional or unintentional). A wireless network can extend beyond the plant perimeter and therefore is open to external agents. WirelessHART defined features to combat these potential attacks. From a network protection point of view, field devices, including instruments cannot provide new encryption keys. Only the *security manager* provides this service..

WirelessHART does not use TCP/IP and so is safe from attacks caused by the ever-evolving internet protocol hacking tools.

In WirelessHART, only *trusted* devices are allowed to participate in the network. Non-trusted communications are seen as coming from interferers; if a message fails to authenticate, then there is no acknowledgement and the received PPDU are thrown away. All the communications between endpoints are handled through private sessions. Only the end devices know session key.

WirelessHART was designed to address all the known wireless security threats. The following sections provide a short summary of how WirelessHART addresses the most common type of attacks.

Denial of Service

Denial of service occurs when the network is flooded—intentionally or unintentionally—with data packets or network commands (e.g., join requests) so that the regular traffic cannot access its destination.

WirelessHART addresses denial of service attacks by limiting the bandwidth available for join request, so that a flooding of this type would have no impact on the network. In addition, packets without the correct MIC are discarded and do not propagate through the network.

Finally, the *network manager* keeps a log of communication statistics and can report anomalous activity and raise alerts for site security to locate an offensive device in the vicinity of the *network devices* being attacked.

Replay Attack

Nonce at network layer is time-slot dependent—so regular replay attacks do not work. In the event of a *replay attack*, the *network manager* reports unusual network activity and directs an alarm to site security.

A *replay attack* on the network layer cannot be reported reliably. The mesh structure can result in duplicating some packets if an acknowledgement was not received, e.g., due to interference. Thus only excessive replay attempts can be reported.

In addition, nonces are also used in the DLL MIC, adding another level of protection against replay attacks on this layer.

Cloned Devices

A clone device uses the credentials of another device to infiltrate the network. To do so, a clone device needs three key pieces of information:

- Unique Identifier (UID) of the device attempting to clone
- A network device TAG
- The join key

A clone device might get access to the UID and the device TAG, but it cannot know the join key, which the security manager might rotate periodically. Even if the security manager does not rotate the join key, the key is a write-only parameter that cannot be retrieved from the device.

SECURITY MANAGER FEATURES

The security manager authenticates new joining devices based upon device unique ID, Network ID, device TAG, and join key. It also authenticates messages to ensure that they have not been altered as they pass through the wireless network.

The WirelessHART security manager is responsible for the generation, storage and management of the encryption keys:

- Join
- Network
- Session

The join key provides a secure method to add a new device to the network. It could be a fixed key or a key that is different for each device added. The end users make the choice on their security procedures.

The join key is not transmitted over the air but by local hard-wired connection from a device with a HART interface. The rationale behind this process is that eavesdropping is prevented.

Session keys ensure that messages are enciphered so that only the final destination can decipher and use the payload created by a source device.

The network key is used as a common network encryption key. The network key is shared among all devices on a network and can be either fixed or rotated to provide a greater level of security. The network key is also used to generate the MIC at the DPDU to authenticate the packet at the DLL level, helping verify the integrity of the packet.

Chapter 13 Concluding Remarks

It has been more than 10 years since the original visionaries of the IEEE Std 802.15.4 met for the first time. The time since has witnessed the creation of a new industry around wireless sensing and actuation, enabling machine-to-machine communications. As originally planned, the ZigBee Alliance organization has been a major platform sponsoring the use of IEEE Std 802.15.4 for residential, commercial, and consumer electronic applications. Similarly, the HART Communication Foundation Communications has served as a major catalyst for WSN applications in the industrial space. In parallel, the IEEE Std 802.15.4 community continues growing with the development of amendments to the original standard that are specialized for particular geographical areas or emphasized in particular classes of applications.

It is difficult to predict the next 10 years in the WSN sector. The growth in this industry will be highly correlated to the success of many major applications, such as smart Grid, industrial process control, and possibly home automation. Most likely, new major sectors, such as building automation, will become prominent.

Another factor is that other wireless technologies are trying to participate in the WSN space. Low-power WiFi and Bluetooth chipsets are tailored to support WSN applications. Today their low-power modes are still much higher than that of existing mature IEEE Std 802.15.4 chipsets but significantly lower power than few years ago. Extrapolating this trend, it is easy to see that unless IEEE Std 802.15.4-based technology accelerates its penetration in its intended markets, we will see interesting dynamics in the market place.

IEEE Std 802.15.4 was designed to meet the needs of an application space requiring only moderate data throughput, demanding very low power consumption and very low implementation cost. The designers of IEEE Std 802.15.4 worked hard to produce a standard that is flexible enough to meet the demands of a wide range of applications—from the wireless sensor networks used in intelligent agriculture and military applications, to the industrial control and monitoring networks, to consumer electronics and home automation uses—while retaining a minimum of costly options and rarely used special features that, while optimizing performance in one application, add complexity to all others. The creators of IEEE Std 802.15.4 believed that striking the proper balance between specialization and complexity maximized the utility of the standard to the wireless communication industry.

The most important strategy for systems architects and decision makers is to always focus their development around customer needs and the related application requirements, and then to pick the right technology that more closely addresses those needs and requirements—do not pick the technology before having a complete definition of the application. In other words, IEEE Std 802.15.4 is not a solution to any problem but an application enabler.

References

[1] Braley, Richard C., Gifford, Ian C., and Heile, Robert F., "Wireless Personal Area Networks: An overview of the IEEE P802.15 working group," *ACM Mobile Computing and Communications Review*, vol. 4, no. 1, January 2000, pp. 26–34.

[2] Callaway, Ed, Bahl, Venkat, Gorday, Paul, Gutiérrez, José A., Hester, Lance, Naeve, Marco, and Heile, Robert, "Home Networking with IEEE 802.15.4, a Developing Standard for Low-Rate Wireless Personal Area Networks," *IEEE Communications Magazine*, special issue on Home Networking, vol. 40, no. 8, August 2002, pp. 70–77.

[3] Craig, William C., "ZigBee: Wireless Control That Simply Works," ZigBee Alliance white paper.

[4] Ennis, G., "Impact of Bluetooth on 802.11 direct sequence," *IEEE P802.11-98/319*, 1998.

[5] European Conference of Postal and Telecommunications Administration (CEPT), European Radiocommunications Committee; Relating to the Use of Short Range Devices (SRD). CEPT/ERC/REC Recommendation 70–03. December 2002.

[6] European Telecommunication Standards Institute, Electromagnetic compatibility and Radio spectrum Matters (ERM); Short Range Devices (SRD); Radio equipment to be used in the 25 MHz to 1 000 MHz frequency range with power levels ranging up to 500 mW; Part 1: Technical characteristics and test methods. ETSI EN 300 220-1 V1.3.1. Sophia-Antipolis, France: European Telecommunication Standards Institute, September 2001.

[7] European Telecommunication Standards Institute, Electromagnetic Compatibility and Radio Spectrum Matters (ERM); Wideband Transmission Systems; Data Transmission Equipment Operating in the 2,4 GHz ISM Band and Using Spread Spectrum Modulation Techniques. ETSI EN 300-328.

Sophia-Antipolis, France: European Telecommunication Standards Institute. 2001.

[8] Gutiérrez, José, Naeve, Marco, Callaway, Ed, Bourgeois, Monique, Mitter, Vinay, and Heile, Robert F., "IEEE 802.15.4—A Developing Standard for Low-Power, Low-Cost Wireless Personal Area Networks," *IEEE Network Magazine*, vol. 15 no. 5, September/October 2001, pp. 12–19.

[9] Heegard, Chris, Coffey, John (Seán) T., Gummadi, Srikanth, Murphy, Peter A., Provencio, Ron, Rossin, Eric J., Schrum, Sid, and Shoemake, Matthew B., "High-performance wireless ethernet," *IEEE Communications*, vol. 39, no. 11, November 2001, pp. 64–73.

[10] Howitt, I., "Bluetooth performance in the presence of 802.11b WLAN," *IEEE Transactions on Vehicular Technology*, vol. 51, 2002.

[11] Howitt, I., "WLAN and WPAN Coexistence in UL Band," *IEEE Transactions on Vehicular Technology*, vol. 50, 2001, pp. 1114-1124.

[12] Howitt, I., and Gutiérrez, J. A., "IEEE 802.15.4 low rate wireless personal area network coexistence issues," *Proceedings of the WCNC 2003*, 2003.

[13] Howitt, I., Mitter, V., and Gutiérrez, J., "Empirical study for IEEE 802.11 and Bluetooth interoperability," *IEEE Spring VTC 2001*, Rhodes, 2001.

[14] IEEE Std 802.11™-1999, IEEE Standard for Information technology—Telecommunications and information exchange between systems—Local and metropolitan area networks—Specific requirements—Part 11: Wireless LAN Medium Access Control (MAC) and Physical Layer (PHY) specifications.

[15] IEEE Std 802.15.1™-2002, IEEE Standard for Information technology—Telecommunications and information exchange between systems—Local and metropolitan area networks—Specific requirements—Part 15.1: Wireless Medium Access Control (MAC) and Physical Layer (PHY) Specifications for Wireless Personal Area Networks (WPANs).

[16] IEEE Std 802.15.4-2003, IEEE Standard for Information technology—Telecommunications and information exchange between systems—Local and

metropolitan area networks—Specific requirements—Part 15.4: Wireless Medium Access Control (MAC) and Physical Layer (PHY) Specifications for Low-Rate Wireless Personal Area Networks (WPANs).

[17] International Electrotechnical Commission, IEC 62591 Ed. 1 International Standard—Industrial communication networks—Wireless communication network and communication profiles —WirelessHART™, April 2010.

[18] Karaouz, Jeyhan, "High-rate Wireless Personal Area Networks," *IEEE Communications*, vol. 39, no. 12, December 2001, pp. 96–102.

[19] Marks, Roger B., Gifford, Ian C., and O'Hara, Bob, "Standards in IEEE 802 unleash the wireless internet," *IEEE Microwave*, vol. 2, no. 2, June 2001, pp. 46–56.

[20] O'Hara, Bob, Petrick, Al, *IEEE 802.11 Handbook: A Designer's Companion*. New York: Standards Information Network, IEEE Press, 1999.

[21] Poor, Robert D., *Embedded networks: Pervasive, low-power, wireless connectivity*, Ph.D dissertation, 2001, Massachusetts Institute of Technology, Cambridge, MA.

[22] Schwetlick, Horst and Wolf, Andreas, "PSSS—Parallel Sequence Spread Spectrum Application in RF communication," Proceedings from the *International Symposium on Signals, Systems, and Electronics (ISSSE)* 2004.

[23] Siep, Tom, *An IEEE Guide: How to Find What You Need in the Bluetooth™ Spec.* New York: Standards Information Network, IEEE Press, 2000.

[24] Siwiak, Kai, *Radiowave Propagation and Antennas for Personal Communications*, 2nd ed., Boston: Artech House.1998.

[25] U.S. Code of Federal Regulations, vol. 47, sec. 15.247. Washington, D.C.: U.S. Government Printing Office. 2001.

[26] U.S. Code of Federal Regulations, vol. 47, sec. 15.249. Washington, D.C.: U.S. Government Printing Office. 2001.

[27] U.S. Department of Commerce, National Institute of Standards and Technology, Specification for the Advanced Encryption Standard (AES).

Federal Information Processing Standards Publication 197. November 26, 2001. http://csrc.nist.gov/publications/fips/fips197/fips-197.pdf.

[28] van Nee, Richard, "A new OFDM standard for high rate wireless LAN in the 5 GHz band," *Proc. IEEE Veh. Tech. Conf.*, 1999, vol. 1, pp. 258–262.

[29] Zimmermann, Hubert, "OSI reference model—The ISO model of architecture for Open Systems Interconnection," *IEEE Trans.Commun.*, vol. COM-28, no. 4, April 1980, pp. 425–432.

[30] Zyren, J., "Reliability of IEEE 802.11 Hi rate DSSS WLANs in a high density Bluetooth environment," *Bluetooth '99*, 1999.

[31] http://www.zigbee.org

[32] http://www.hartcomm.org

[33] http://www.isa.org

[34] http://www.ietf.org/html.charters/6lowpan-charter.html

Glossary

access control list (ACL): A table used by a device to determine which devices are authorized to perform a specific function.

ad hoc network: A wireless network composed of communicating devices without preexisting infrastructure. An ad hoc network is typically created in a spontaneous manner and is self-organizing and self-maintaining.

adapter: A device that provides physical and logical connectivity between two distinct communication protocols.

advanced encryption standard (AES): The symmetric encryption algorithm specified by the U.S. government in Federal Information Processing Standard (FIPS) 197. (http://csrc.nist.gov/encryption/aes/)

application layer: The top layer in the open system interconnection (OSI) data communication model, in which application functions are performed that directly affect the end user.

association: The service used to establish a device's membership in a wireless personal area network.

authentication: The service used to establish the identity of one device as a member of the set of devices authorized to communicate securely to other devices in the set.

authentic data: Data with a source verifiable through cryptographic means.

bandwidth: The difference between the highest and lowest frequencies employed in a communication channel. More informally, the largest amount of data that can be sent per unit time over a communication channel.

baseband: A communication channel through which constant-potential symbols (e.g., dc voltages) are sent, without frequency shifting (e.g., modulation of a high-frequency carrier); the band of frequencies occupied by the signal before it modulates the carrier.

baud rate: A measure of signaling speed equal to the number of discrete conditions or signal events per second; the number of symbols per second in a communication system.

bit rate: A measure of the amount of information per second in a communication system. Also, the number of binary symbols per second in a communication system.

Bluetooth: A Wireless Personal Area Network specification sponsored by the Bluetooth Special Interest Group (http://www.bluetooth.org) and standardized as IEEE Std 802.15.1.

broadband: Having a large bandwidth when compared to a reference; especially, a modulated signal having a large bandwidth when compared to its associated baseband signal. More informally, a high bit rate suitable for multimedia applications.

channel hopping: Technique used to provide frequency diversity to DSSS systems. This technology helps mitigate in-band interference and reduce multi-path fading effects.

clear channel assessment: An evaluation of the communication channel prior to a transmission to determine if the channel is occupied.

confidentiality: Assurance that communicated data remains private to the parties for whom the data are intended.

coordinator: A full function device (FFD) with network device functionality that is also capable of providing synchronization services through the transmission of beacons and the creation of a superframe structure. If a coordinator is the principal controller of a PAN, it is called the PAN coordinator.

coverage area: The area where two or more IEEE Std 802.15.4 units can exchange messages with acceptable quality and performance.

cyclic redundancy check (CRC): An error detection code transmitted with a block of data in order to detect corruption of the data.

data integrity: Assurance that received data have not been modified from their original form.

data link layer: The layer between the Physical and Network layers of the open system interconnection (OSI) reference model for communication between computer systems, providing the functional and procedural means to establish, maintain, and release data links between network entities.

disassociation: The service that removes an existing association.

Electronic Device Description Language (EDDL): A text-based, domain specific language that is normally used in the industrial applications to describe the characteristics of field devices.

frame: The format of aggregated bits from a MAC sublayer entity that are transmitted together in time.

full function device (FFD): A device capable of operating as a coordinator or network device, implementing the complete protocol set.

gateway: A device that connects the wireless network to a backhaul network, allowing data to flow between the two networks.

IEEE: *See:* **Institute of Electrical and Electronic Engineers**

IEEE Std 802.11: IEEE standard that specifies medium access and physical-layer specification for wireless connectivity among fixed, portable, and moving devices within a local area at data rates of 1 and 2 Mb/s.

IEEE Std 802.11a: Extension of IEEE Std 802.11 using orthogonal frequency-division multiplexing (OFDM) modulation, operating in the 5GHz frequency band at data rates of up to 54 Mb/s.

IEEE Std 802.11b: Extension of IEEE Std 802.11 using Direct Sequence Spread Spectrum modulation in the 2.4 GHz frequency band at data rates of up to 11 Mb/s.

industrial, scientific, and medical (ISM) bands: Radio-frequency bands reserved for these applications by international agreement. The ISM bands relevant to WPANs are located at 900 MHz, 2.4 GHz, and 5.7 GHz.

Institute of Electrical and Electronic Engineers (IEEE): A non-profit, technical professional association with the goal of advancing global prosperity by fostering technological innovation, enabling members' careers, and promoting community worldwide. IEEE promotes the engineering process of creating, developing, integrating, sharing, and applying knowledge about electrical and information technologies and sciences for the benefit of humanity and the profession (http://www.ieee.org).

integrity code: A data string generated using a symmetric key that is typically appended to data in order to provide data integrity and source authentication (also called a message integrity code).

International Electrotechnical Commission: The International Electrotechnical Commission (IEC) is a global organization that prepares and publishes international standards for all electrical, electronic and related technologies.

International Organization for Standardization (ISO): A worldwide, non-governmental federation of national standards bodies, promoting the development of standardization and related activities in the world with a view to facilitating the international exchange of goods and services, and to developing cooperation in the spheres of intellectual, scientific, technological and economic activity (http://www.iso.org).

International Telecommunications Union (ITU): International organization within the United Nations system in which governments and the private sector coordinate global telecommunications networks and services (http://www.itu.int).

ISO: *See:* **International Organization for Standardization**

ITU: *See:* **International Telecommunications Union**

key establishment: A process by which two entities securely establish a symmetric key that is known only by the participating entities.

key management: Methods for controlling keying material throughout its life cycle, including creation, distribution, and destruction.

key transport: A process by which an entity sends a key to another entity.

ISM bands: *See:* **industrial, scientific, and medical bands**

logical channel: One of a number of distinct channels on a physical communication link.

Logical Link Control sublayer: The upper portion of the Data Link Layer in the OSI reference model, responsible for the organization of data flow.

low-rate wireless personal area network (LR-WPAN): A wireless personal area network, optimized for low-data-rate applications emphasizing long battery life and low implementation cost.

MAC sublayer: *See:* **Medium Access Control sublayer**

MAC protocol data unit (MPDU): The unit of data exchanged between two Medium Access Control entities.

Medium Access Control sublayer (MAC sublayer): The lower portion of the Data Link Layer in the open system interconnection (OSI) reference model, responsible for acquiring the right to use the underlying physical communication medium.

message integrity code: *See:* **integrity code**

mobile device: A device that uses network communications while in motion.

narrowband: Having a small bandwidth when compared to a reference, especially, a modulated signal having a bandwidth comparable to, or less than, its associated baseband signal. More informally, a low bit rate suitable for voice and data applications.

network access point: The device that provides the interconnection between the wireless network and the gateway.

network device: A reduced function device (RFD) or full function device (FFD) implementation containing an IEEE Std 802.15.4 medium access control and physical interface to the wireless medium.

network layer: The layer between the Data Link and Transport layers of the open system interconnection (OSI) reference model for communication between computer systems, providing routing and switching functions and procedures.

network manager: A device dedicated to configuring and optimizing the WirelessHART network.

network topology: The logical structure of a network.

Open System Interconnection (OSI): The seven-layer ISO model of architecture for communication between (possibly heterogeneous) computer systems.

orphaned device: A device that has lost contact to its associated personal area network coordinator.

packet: The format of aggregated bits that are transmitted together in time across the physical medium.

PAN coordinator: A coordinator that is the principal controller of a personal area network (PAN). An IEEE Std 802.15.4 network has exactly one PAN coordinator.

parent: The PAN coordinator or coordinator through which a network device has associated with.

payload data: The contents of a data message that is being transmitted.

payload protection: The generic term for the provision of security services on payload data, including confidentiality, data integrity, and authentication.

peer-to-peer network: A substantially homogeneous network in which devices communicate as equals, often employing distributed, multihop routing protocols.

personal area network (PAN): A wireless network for the personal operating space.

personal operating space: The space about a person or object that is typically about 10 m in all directions and envelops the person or object whether stationary or in motion.

physical layer: The layer below the Data Link Layer of the open system interconnection (OSI) reference model for communication between computer systems, providing the physical connections between data link entities.

portable device: A device that may be moved from location to location, but only uses network communications while at a fixed location.

protocol data unit (PDU): The unit of data exchanged between two peer entities.

pseudo-random number generation: The process of generating a deterministic sequence of bits from a given seed that has the statistical properties of a random sequence of bits when the seed is not known.

random number generator: A device that provides a sequence of bits that is unpredictable.

reduced function device (RFD): A device operating with a minimal implementation of IEEE Std 802.15.4 protocol that is capable only of functioning as a network device.

security manager: A device dedicated to the generation, storage, distribution, and management of encryption keys in a WirelessHART network.

security suite: A group of security operations designed to provide security services on MAC frames.

service access point: Any entity that provides access to the service interface.

service data unit (SDU): Information that is delivered as a unit through a service access point.

service primitive: The basic unit of service provided between layers of a communication protocol.

star network: A network employing a single, central device through which all communication between devices must pass.

symmetric key: A secret key that is shared between two or more parties that may be used for encryption/decryption or integrity protection/integrity verification depending on its intended use.

time division multiplexing (TDM): A technology in which two or more signals or bit streams are multiplexed as subchannels in one communication channel by dividing it into recurrent timeslots of fixed length, one for each subchannel.

transaction: The exchange of related, consecutive frames between two peer medium access control (MAC) entities, required for a successful transmission of a MAC command frame or a data frame.

transport layer: The layer between the Network and Session layers of the open system interconnection (OSI) reference model for communication between computer systems, providing transparent transfer of data between session entities.

wireless local area network (WLAN): A computer communication network spanning at most a campus-sized (~1 km) area, employing radio, infrared, or other wireless physical media.

wireless medium (WM): The medium used to implement the transfer of protocol data units (PDUs) between peer physical layer (PHY) entities of a low-rate wireless personal area network (LR-WPAN).

wireless metropolitan area network (WMAN): A computer communication network spanning at most a city-sized area, employing radio, infrared, or other wireless physical media.

wireless personal area network (WPAN): A computer communication network spanning the personal operating space, employing radio, infrared, or other wireless physical media.

ZigBee Alliance: An association of companies working together to create a very low-cost, very low power consumption, two-way, wireless communications standard (http://zigbee.org).

Index

modulation 31
baud rate 232
beacon 124–125, 232
 frame 40, 105–115
 management 105–108
 payload 111, 116, 131
beacon packet 30
beacon request command 97, 208
Bellman-Ford algorithm 127, 130
Binary Phase Shift Keying (BPSK)
 See BPSK
bit rate 232
blocking 72, 172
Bluetooth xxiii, 4, 15, 24, 160, 225, 232
BPSK xxv, xxvi, 31, 45–54, 55, 68
 mode specification 58–60

C

cable replacement 4, 15
carrier sense multiple access with
 collision avoidance (CSMA/CA) 79
carrier sense multiple access-collision
 avoidance (CSMA-CA) 37, 84, 87,
 101, 113, 208
channel
 access 34, 37–38, 125
 assignment 49–53
 capacity 37
 filters 32
 hopping xxvi, 171, 172, 174, 232
 scan 159
chip 31, 55, 56, 145–146
 single 44
 system-on-a-chip
 See system-on-a-chip (SoC)
cipher, symmetric 116
ciphertext 116
clear channel assessment (CCA) xxix,
 43, 75, 172, 232
cluster 119, 130–134
cluster head 36, 82
cluster-tree 36–42, 82, 119, 135–138
coexistence 12, 33, 161, 172–173

coherence 32
collision 37, 41, 79, 170, 173, 174, 209
colocated transceivers 161–162
commands
 common practice 168
 device specific 168
 universal 168
communication range 10
confidentiality 232
confirm primitive 73, 102, 120, 122
consumer electronics 20, 21, 226
Consumer Electronics Association
 (CEA) 19
contention access period (CAP) xxix,
 75, 84, 87, 88, 101, 102, 103, 115
contention-based channel access 125
contention-free period xxix, 85, 101,
 103, 115
coordinator
 PAN coordinator 236
coordinator realignment command 104,
 105, 114
cost considerations
 data rates 46
coverage area 232
cyclic redundancy check (CRC) xxix,
 39, 110, 232

D

data frame 40, 41, 87–88, 89, 94, 99,
 111, 113, 125, 238
data integrity 232
data link layer (DLL) xxix, 6, 79, 186,
 207, 233, 235, 236
data packet 74, 172, 173, 176, 209
data request command 105, 108, 113,
 208
data transmission 92, 173, 174
despreading 32
device specific commands
 See commands, device specific
device synchronization 81, 82, 97, 105

peer-to-peer 36, 80, 82, 87, 94, 106,
119, 125, 126
self-healing 190
star ??–81, 82–??, 120, 237
topology 236
network access point 174, 177–181, 235
network manager 171, 174–183, 187–
190, 235

O

OEM
See original equipment manufacturer
(OEM)
open system interconnection (OSI) 236
O-QPSK xxv, xxvi, xxxi, 31, 45
mode specification 60–62
original equipment manufacturer
(OEM) 12, 58, 86, 146–??
orphan device management 104
orphan notification command 104
orphaned device 236
OSI
See open system interconnection
(OSI)

P

packet 236
PAN 236
applications 15–26
ID conflict 105, 114
maintaining 114
starting 114
PAN coordinator 34–35, 36, 81–87, 90,
97–114, 115, 119–136, 232, 236
PAN information base 74, 95–96
parallel sequence spread spectrum
(PSSS)
See spread spectrum, parallel
sequence (PSSS)
parent 236
parent-child relationship 37
payload 40, 41, 236
data 41, 78, 116
MAC command 113

payload protection 236
PC peripherals 20–21
PC-enhanced 23
PDA
See personal data assistant (PDA)
PDU
See protocol data unit (PDU)
peer-to-peer
See network, peer-to-peer
personal data assistant (PDA) 15, 17
personal healthcare 22–23
personal operating space (POS) 4, 236
physical layer (PHY) 43–78, 236
bandwidth 45, 53, 72, 84, 85, 99, 103
data rates 45–49
data service 54, 58, 60, 62, 73
header 63
management service 73–74
market considerations 38
regulatory considerations 43–46
specifications xxvi
synchronization header 66
physical layer service data unit
(PSDU) 40, 78
PIB
See PAN information base
plaintext 116
PLME 73
PLME-CCA 76
PLME-ED 76, 97
PLME-GET 74
PLME-SAP 73
PLME-SET 74
PLME-SET-TRX-STATE 75
polling 82, 108
portable device 236
power consumption 238
power cycling 10
power spectral density 32, 33, 68
precision agriculture 25–26
processing gain 32, 141
PROFIBUS 184, 189
propagation xxiv, 8, 12, 70, 187, 190

protocol 91, 126
protocol data unit (PDU) 237
protocol data units 92
protocol layer 40, 77, 78, 110
protocol stack 37, 79
pseudo-random number generation
237
pseudo-random sequence 55, 58
PSSS
  See spread spectrum, parallel
    sequence (PSSS)

Q

quality of service (QoS) 3, 4, 34, 38, 39,
85, 189, 190
quasi-orthogonal 55, 56, 60, 66

R

radio frequency (RF) 22, 97, 157
radio frequency identification (RFID) 9
random number generator 237
range 9, 11
realignment 104, 105, 114
receiver selectivity 72
reduced function device (RFD) 35, 125,
235, 237
regulation 44, 47
regulations 8, 11, 43
regulatory agencies 12, 38
remote keyless entry 24
replay attacks
  See security, replay attack
request primitive 97–100, 111–115,
120–123
response primitive 73, 94, 99, 108
routing table 13, 82, 127, 130, 133–136,
176, 182

S

secured mode
  See security, secured mode
security
  access control xxv, 79
  access control list (ACL) 231

advanced encryption standard
  (AES) 118, 231
encryption 116, 182, 183, 219–224,
  237
information 165
message integrity code (MIC) 116,
  117
replay attack 117, 223
secured mode 117
services 115–118
suites 237
unsecured mode 117
security manager 177–183, 189, 223,
237
security suites 118
self-healing 82, 187
  See network, self-healing
self-organization 13, 199, 231
sensitivity 34, 68
sequence number 110, 112
service access point 237
service data unit (SDU) 237
service primitive 237
short-range 3, 9, 10, 12, 15, 29
signal-to-noise ratio (SNR) 32
single chip
  See chip, single
SoC
  See system-on-a-chip (SoC)
spread spectrum 11
  frequency hopping (FHSS) 141
  parallel sequence (PSSS) xxv
spread spectrum, parallel sequence
  (PSSS) 31, 45–55, 63–66, 77
standby power 30, 31, 150
star network
  See network, star
stick-on sensor 15, 35
superframe 30, 40, 41, 80, 86–89, 97,
  100, 103, 111, 115, 125, 176, 177, 183,
  208
  boundary marker 40
  structure 83–86

symbol to chip mapping 56, 62, 67
symmetric key 237
synchronization 141, 232
  beaconless 108
  control 105
  device 81, 115
  frame 41
  header (SHR) 40, 77
  time 173, 190, 210
synthesizer 142
system-on-a-chip (SoC) 33, 158

T
TDMA
  See time division multiplexing
    (TDMA)
telematics 24
time division multiplexing
  (TDMA) xxiv, xxvi, 170, 183, 237
time-slotted channel hopping
  (TSCH) xxiv, 171
tire pressure monitoring 15, 25
toys and games 19, 23–24
transaction 238
transmission scenario 93
transmitted noise 161
transmitter 11, 30–32, 70, 144, 148
  cellular 162
  oscillator 143
  power 33, 68
  radio 75
  spread spectrum 33
  wireless 77
transport layer 238
TSCH
  See time-slotted channel hopping
    (TSCH)

type approval 43

U
ubiquitous 29
ultrasound 9
universal commands
  See commands, universal
unlicensed bands 11, 12, 29, 33, 38, 43,
  44–46, 79
unsecured mode
  See security, unsecured mode
untethered 10, 15, 29

V
virtual wire 15, 25

W
wideband FM 31
wireless hub 15, 17
wireless medium (WM) 238
wireless metropolitan area network
  (WMAN)
  See WMAN
wireless sensor network (WSN) 7–13
WirelessHART 165–224
  coexistence 172–173
  network reliability 173–183
WLAN 4–5, 46, 79, 141, 144, 238
WM
  See wireless medium (WM)
WMAN 4, 5
WPAN xxvii, 4–6, 46, 72, 79, 157, 161,
  162, 233, 238

Z
ZigBee Alliance xxvii, 6, 20, 23, 238

CPSIA information can be obtained at www.ICGtesting.com
Printed in the USA
BVOW08s1927240215

389135BV00005B/13/P